D0948837

Graduate Texts in Contemporary Physics

Graduate Texts in Contemporary Physics

Assa Auerbach

Interacting Electrons and Quantum Magnetism

With 34 Illustrations

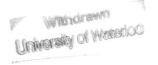

Springer-Verlag
New York Berlin Heidelberg London Paris
Tokyo Hong Kong Barcelona Budapest

Assa Auerbach
Technion
Israel Institute of Technology
Department of Physics
Haifa 32000, Israel

Series Editors
Joseph L. Birman
Department of Physics
The City College of the
  City University of New York
New York, NY 10031
USA

Helmut Faissner
Physikalisches Institut
RWTH Aachen
5100 Aachen, Germany

Jeffrey W. Lynn
Department of Physics and Astronomy
University of Maryland
College Park, MD 20742
USA

Library of Congress Cataloging-in-Publication Data
Auerbach, Assa.
    Interacting electrons and quantum magnetism / Assa
Auerbach.
      p. cm.
    Includes bibliographical references and index.
    ISBN 0-387-94286-6 (New York). — ISBN 3-540-94286-6 (Berlin)
    1. Energy-band theory of solids.   2. Wave functions.
3. Integrals, Path.   4. Electron-electron interactions.
5. Magnetism.   I. Title.
QC176.8.E4A94   1994
530.4′12 — dc20   94-6510

With illustrations by Dick Codor.

Printed on acid-free paper.

Production managed by Hal Henglein; manufacturing supervised by Jacqui Ashri.
Photocomposed copy prepared from the author's LaTeX files.
Printed and bound by R.R. Donnelley & Sons, Harrisonburg, VA.
Printed in the United States of America.

9 8 7 6 5 4 3 2 1

ISBN 0-387-94286-6 Springer-Verlag New York Berlin Heidelberg
ISBN 3-540-94286-6 Springer-Verlag Berlin Heidelberg New York

To my parents, Ruth and Israel.

Preface

In the excitement and rapid pace of developments, writing pedagogical texts has low priority for most researchers. However, in transforming my lecture notes[1] into this book, I found a personal benefit: the organization of what I understand in a (hopefully simple) logical sequence. Very little in this text is my original contribution. Most of the knowledge was collected from the research literature. Some was acquired by conversations with colleagues; a kind of physics oral tradition passed between disciples of a similar faith.

For many years, diagramatic perturbation theory has been the major theoretical tool for treating interactions in metals, semiconductors, itinerant magnets, and superconductors. It is in essence a weak coupling expansion about free quasiparticles. Many experimental discoveries during the last decade, including heavy fermions, fractional quantum Hall effect, high-temperature superconductivity, and quantum spin chains, are not readily accessible from the weak coupling point of view. Therefore, recent years have seen vigorous development of alternative, nonperturbative tools for handling *strong* electron–electron interactions.

I concentrate on two basic paradigms of strongly interacting (or constrained) quantum systems: the Hubbard model and the Heisenberg model. These models are vehicles for fundamental concepts, such as effective Hamiltonians, variational ground states, spontaneous symmetry breaking, and quantum disorder. In addition, they are used as test grounds for various nonperturbative approximation schemes that have found applications in diverse areas of theoretical physics.

The level of the text should be appropriate for a graduate student with some background in solid state physics (single electron theory) and familiarity with second quantization. The exercises vary in difficulty and complement the text with specific examples and corollaries. Some of the mathematical background material is relegated to the appendices.

I owe most to the relentless efforts of Maxim Raykin, whose careful proofreading weeded out inconsistencies and helped clarify numerous points. I am also heavily indebted to Duncan Haldane, who introduced me to quantum magnetism, and to my friend and colleague, Dan Arovas, who taught me about parent Hamiltonians, the single mode approximation, and many

---

[1]for a graduate course on Quantum Many Particle Systems given at Boston University and at the Technion during 1990–1993.

other things,[2] and for his critical comments. I am grateful for the support of the Alfred P. Sloan Foundation, which enabled me to complete this book.

Assa Auerbach
Haifa, 1994

---

[2]including the use of *phantom daggers.*

Contents

Part I

Basic Models

Illustration by Dick Codor.

1

Electron Interactions in Solids

1.1 Single Electron Theory

A single electron moving in a periodic potential is described by the band structure equation

$$H^0 \phi_{\mathbf{k}s}(\mathbf{x}) \equiv \left[-\frac{\hbar^2}{2m} \nabla^2 + V^{ion}(\mathbf{x}) \right] \phi_{\mathbf{k}s}(\mathbf{x}) = \epsilon_{\mathbf{k}} \phi_{\mathbf{k}s}(\mathbf{x}), \qquad (1.1)$$

where $\phi_{\mathbf{k}s}$ and $\epsilon_{\mathbf{k}}$ are the Bloch wave function and band energy, respectively. $\mathbf{k}$ is the electron's lattice momentum, and $s = \uparrow, \downarrow$ is its spin in the S^z direction. Here we supress the band index and ignore spin orbit coupling.

For N_e electrons, the Schrödinger equation can be reduced to a set of band structure equations (1.1) only if the Hamiltonian is a separable sum of single particle Hamiltonians,

$$\mathcal{H}^0 = \sum_{i=1}^{N_e} H^0[\nabla_i, \mathbf{x}_i]. \qquad (1.2)$$

The eigenstates of (1.2) are *Fock states* (see Appendix A) constructed by Slater determinants:

$$\Psi_{[\mathbf{k},s]}^{Fock}(\mathbf{x}_1, \ldots, \mathbf{x}_{N_e}) = \det_{ij} \left[\phi_{\mathbf{k}_i, s_i}(\mathbf{x}_j) \right] \qquad (1.3)$$

and the eigenenergies are

$$E_{[\mathbf{k}]} = \sum_{i=1}^{N_e} \epsilon_{\mathbf{k}_i}. \qquad (1.4)$$

In the ground state the lowest N_e states are occupied, and the uppermost energy is called the *Fermi energy*,

$$\max \epsilon_{\mathbf{k}} = \epsilon_F. \qquad (1.5)$$

Equation (1.5) defines the *Fermi surface* in $\mathbf{k}$ space.

The Hamiltonian that includes interactions between electrons is

$$\mathcal{H} = \mathcal{H}^0 + \frac{1}{2} \sum_{i \neq j} v^{el-el}(\mathbf{x}_i, \mathbf{x}_j). \qquad (1.6)$$

v^{el-el} spoils the separability of (1.2) and makes the $\mathcal{H}$ much harder to diagonalize than $\mathcal{H}^0$.

Much of the Coulomb interaction effects can be incorporated into the single particle part of $\mathcal{H}$ by modifying the ion potential

$$\mathcal{H}^0 \to \tilde{\mathcal{H}}^0 = \sum_{i=1}^{N_e} \left(H^0[\nabla_i, \mathbf{x}_i] + V^{ion}(\mathbf{x}_i) + v^{eff}[\mathbf{x}_i, \rho] \right), \qquad (1.7)$$

where v^{eff} is a functional of the ground state density $\rho(\mathbf{x})$. There are various approximation schemes for $v^{eff}[\rho]$ which are reviewed in the literature (see bibliography). Most band structure calculations for specific materials take v^{eff} to be local, i.e., to depend on ρ (and perhaps its spatial derivatives) at $\mathbf{x}$. ρ, in turn, is determined self-consistently by solving for the ground state of (1.7).

The residual interactions are

$$\tilde{v}_{ij} = v^{el-el}(\mathbf{x}_i, \mathbf{x}_j) - \left[v^{eff}(\mathbf{x}_i) + v^{eff}(\mathbf{x}_j) \right] / N_e. \qquad (1.8)$$

The transformation $v^{el-el} \to \tilde{v}$ represents *screening*. In reality, screening is a dynamical process which involves collective charge fluctuations with a *plasma frequency* scale. If the plasma frequency is higher than excitation energies of interest, $\tilde{v}$ can be taken to be instantaneous. In metals, the screened interaction decays exponentially within a Thomas–Fermi screening length λ^{TF}.

Single electron approximation schemes, which ignore $\tilde{v}_{ij}$, have enjoyed a large amount of success in predicting bulk energies and structures of many materials. However, for systems that exhibit magnetism or superconductivity, the residual interactions are crucial.

The effects of $\tilde{v}_{ij}$ can be calculated perturbatively by Feynman diagrams and time-ordered Green's functions. Resummation of classes of diagrams (such as in random phase approximations) may be used to describe instabilities of the noninteracting system towards ordered ground states. These methods are covered in detail in many textbooks, some of which are listed in the bibliography. The techniques that will be emphasized in this book are specifically designed to complement perturbation theory and to treat strong electron–electron interactions.

1.2   Fields and Interactions

In the many electron Hilbert space, it is convenient to use second quantized operators which enforce the antisymmetry of all states (see Appendix A). The *field operator* $\psi_s^\dagger(\mathbf{x})$ creates a particle localized at $\mathbf{x}$ in a spin state s:

$$\langle \mathbf{x}'s' | \psi_s^\dagger(\mathbf{x}) | 0 \rangle = \delta_{ss'} \delta(\mathbf{x} - \mathbf{x}'). \qquad (1.9)$$

Using (A.10) we can expand the field operator in terms of *any* orthonormal single particle basis $\{\phi\}$:

$$\psi_s^\dagger(\mathbf{x}) = \sum_i \phi_i^*(\mathbf{x})c_{is}^\dagger, \tag{1.10}$$

where $c^\dagger, c$ are anticommuting creation and annihilation operators. The field operators therefore obey

$$\{\psi_s^\dagger(\mathbf{x}), \psi_{s'}(\mathbf{x}')\} = \delta(\mathbf{x} - \mathbf{x}')\delta_{ss'}. \tag{1.11}$$

The *local density* operator is defined as

$$\hat{\rho}(\mathbf{x}) = \sum_s \psi_s^\dagger(\mathbf{x})\psi_s(\mathbf{x}) \tag{1.12}$$

$$= \sum_{ii's} \phi_i^*(\mathbf{x})\phi_{i'}(\mathbf{x})c_{is}^\dagger c_{i's}. \tag{1.13}$$

$\hat{\rho}$ measures the probability density of finding a particle of either spin at position $\mathbf{x}$. Note that the local density operator has, in general, off-diagonal terms $i \neq i'$. $\hat{\rho}(\mathbf{x})$ is the second quantized representation of the Schrödinger density operator, whose expectation value in coordinate space is

$$\rho(\mathbf{x}) = \sum_i \delta(\mathbf{x} - \mathbf{x}_i) . \tag{1.14}$$

The Fock state specified by $|\mathbf{x}_1, \mathbf{x}_2, \ldots\rangle$ is an eigenstate of $\hat{\rho}(\mathbf{x})$ with the eigenvalue $\rho(\mathbf{x})$. Therefore it is natural to represent the local operator $\mathcal{H}$ using the local field and density operators (1.10) and (1.13). Separating the single particle and the residual interactions parts as in (1.7 and 1.8) we write

$$\mathcal{H} = \tilde{\mathcal{H}}^0 + \tilde{\mathcal{V}}^{el-el}, \tag{1.15}$$

where

$$\tilde{\mathcal{H}}^0 = \sum_s \int d^3x\, \psi_s^\dagger(\mathbf{x}) \left[-\frac{\hbar^2}{2m}\nabla^2 + V^{ion} + v^{eff}(\mathbf{x}) \right] \psi_s(\mathbf{x}), \tag{1.16}$$

and

$$\tilde{\mathcal{V}}^{el-el} = \frac{1}{2} \int d^3x\, d^3y\, \tilde{v}(\mathbf{x}, \mathbf{y}) \left[\hat{\rho}(\mathbf{x})\hat{\rho}(\mathbf{y}) - \delta(\mathbf{x} - \mathbf{y})\hat{\rho}(\mathbf{x}) \right]$$

$$= \frac{1}{2} \int d^3x\, d^3y\, \tilde{v}(\mathbf{x}, \mathbf{y}) \sum_{ss'} \psi_s^\dagger(\mathbf{x})\psi_{s'}^\dagger(\mathbf{y})\psi_{s'}(\mathbf{y})\psi_s(\mathbf{x}).$$

$$\tag{1.17}$$

In the second line of (1.17) the field operators are normal ordered, which eliminates the self-interaction term $\tilde{v}(0)\delta(\mathbf{x} - \mathbf{y})\hat{\rho}$.

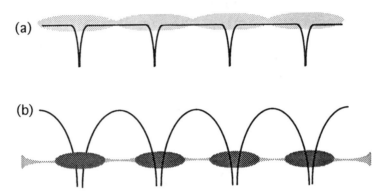

FIGURE 1.1. Conduction electron density for (a) s and p electrons and (b) d and f electrons. Solid lines are effective ionic potentials.

1.3   Magnitude of Interactions in Metals

It must be emphasized that it is not easy to ascertain a priori whether the residual interactions in a given system are to be considered "weak" or "strong." The appropriate question to ask is whether their effect on the ground state correlations and low-lying excitations is dramatic or not. Perturbation theory can serve as a guide to estimate interaction effects. Crudely speaking, the dimensionless parameter for the Coulomb interaction effects in a metal is

$$g = \frac{e^2}{\kappa \epsilon_F \langle r_{ij} \rangle} \tag{1.18}$$

where e is the electron charge, κ is the static dielectric constant, $\langle r_{ij} \rangle$ is average interelectron distance, and ϵ_F is the fermi energy measured from the bottom of the conduction band. As the ions that contribute the conduction electrons are lower in the Periodic Table, the relative interaction strengths increase. The effect of the ionic potential on the s and p conduction electrons is weak. Therefore, the conduction band wave functions are delocalized and the density is quite uniform throughout the crystal, as illustrated in Fig. 1.1(a). As a consequence, interactions between electrons are relatively weak.

In contrast, transition metals and mixed valence rare earth compounds contribute d and f electrons to the conduction band. There, the electrons are mostly localized within a small radius $\langle r_{ij} \rangle \ll a$ around the ions, where a is the lattice spacing. At the same time, the Fermi energy ϵ_F is reduced by large interatomic potential barriers, as shown in Fig. 1.1(b). Thus, one expects transition metals and mixed valence rare earth compounds to have large values of g, and to be considered strongly interacting electron systems.

This classification scheme is only a rough rule of thumb. It cannot predict whether a particular compound is a paramagnetic metal, or has magnetic, charge density or superconducting order. Obviously, a noninteracting degenerate ground state is highly susceptible to degeneracy lifting perturbations. Therefore, the ground state depends sensitively on the details of the band structure and near degeneracies in $\tilde{\mathcal{H}}^0$.

1.4 Effective Models

The quartic interactions in $\tilde{\mathcal{V}}^{el-el}$ make $\mathcal{H}$ a truly many-particle problem and therefore very hard to diagonalize. Numerical methods are restricted to very small systems, since the Hilbert space size grows *exponentially* with the number of electrons and the size of the single-particle basis set. Theorists resort to simplified models (see illustration on title page of Part I), where only a reduced set of single-particle states and interaction matrix elements are included.

The replacement of $\mathcal{H}$ by an effective model can be put on firmer footing when formulated as a renormalization group transformation.[1] When one is interested in the low-frequency and low-temperature correlations, the high-energy states can be eliminated by projecting the Green's function onto the low-energy sector. This results in an effective Hamiltonian $\mathcal{H}^{eff}$ which acts within the lower energies subspace.[2] In a path integral formulation, the high-frequency modes can be integrated out, leaving us with an effective Lagrangian for the low-frequency modes.[3]

The transformation $\mathcal{H} \rightarrow \mathcal{H}^{eff}$ is called *renormalization*. Under renormalization, certain interactions (called *irrelevant*) are suppressed relative to others. The interactions that grow or stay constant are *relevant* or *marginal*, respectively. Thus many of the microscopic details of the band structure and interactions drop out of the effective Hamiltonian which includes only the most relevant interactions. If renormalization results in a model with fewer single-particle states and interactions, it goes a long way toward obtaining the low-energy correlations. The parameters of the effective model could be determined from the microscopic Hamiltonian by solving for the renormalization group flow. Admittedly, such calculations have rarely been done for real materials, and model partameters have been determined by fitting experiments. In the absence of reliable calculations, the relevancy

---

[1]See Shankar's review.

[2]For example: in Section 3.2 the $t - J$ model is derived as the effective Hamiltonian of the large-U Hubbard model.

[3]See for example the renormalization of the nonlinear sigma model in Chapter 13.

or irrelevancy of certain interactions is a source of unsettled controversies between proponents of different models.

One of the minimal models of interacting electrons is the *Hubbard model*, which will be described in Chapter 3. However, even the Hubbard model has only been solved in one dimension. A simpler model, which still includes many-particle interactions, is the quantum Heisenberg model, which is derived in Chapter 3 as a special limit of the Hubbard model. Although it describes only spin degrees of freedom, its ground state and excitations are highly correlated (i.e. far from a combination of a few Fock states). The rich physical properties of the Heisenberg model come under the name "Quantum Magnetism," which is a major subject of this book. We shall devote considerable discussion to spin interactions and their effect on quantum and thermal fluctuations. In a broader context, the quantum Heisenberg model will be used to demonstrate fundamental physical concepts. As a pedagogical tool, we shall test on it various mathematical techniques and approximation schemes that are common to other quantum many-particle systems.

1.5   Exercises

1. Find the total number $N = \partial F/\partial \mu$, entropy $S = -\partial F/\partial T$, and specific heat $C_v = T\partial S/\partial T$ of free bosons and free fermions using (A.24) of Appendix A.

Bibliography

For general background on the electronic structure of solids, there are numerous excellent textbooks, for example:

- N.W. Ashcroft and N.D. Mermin, *Solid State Physics* (Holt, Rinehart, and Winston, 1976);
- C. Kittel, *Quantum Theory of Solids* (John Wiley and Sons, 1987);
- J. Callaway, *Quantum Theory of the Solid State* (Academic Press, 1992);
- W. Jones and N.H. March, *Theoretical Solid State Physics* (Dover, 1973).

The basic techniques of many-particle diagramatic perturbation theory for condensed matter are given by, e.g.:

- A. Fetter and J.D. Walecka, *Quantum Theory of Many Particle Systems* (McGraw-Hill, 1971);
- G. Mahan, *Many Particle Physics* (Pienum Press, 1981);

A recent book by

- P. Fulde, *Electron Correlations in Molecules and Solids* (Springer-Verlag, 1991);

is also recommended as a contemporary theoretical and experimental reference.

Suggested further background reading:

- P.W. Anderson, *Concepts in Solids* (Addison-Wesley, 1963).

An excellent recent tutorial on the renormalization group for interacting electrons is

- R. Shankar, Rev. Mod. Phys. 66, 129 (1994).

2

Spin Exchange

Ferromagnetism is obtained in solids when the magnetic moments of many electrons align. Antiferromagnetism and spin density waves describe oscillatory ordering of magnetic moments. The classical dipolar interaction between the electron moments (which is of order 10^{-5}eV) is far too weak to explain the observed magnetic transition temperatures (which are of order 10^2–10^3 °K in transition metal and rare earth compounds).

It was therefore realized in the early days of quantum mechanics that the coupling mechanism that gives rise to magnetism derives from the following fundamental properties of electrons:

1. The electron's spin.

2. The electron's kinetic (delocalization) energy.

3. Pauli exclusion principle (Fermi statistics).

4. Coulomb repulsion between electrons.

As an introductory example, we study two electrons that are spatially localized by an external potential and repel each other via the Coulomb interaction. We shall find that couplings between the spins of the two electrons can be either ferromagnetic or antiferromagnetic, depending on the nature of the noninteracting states. These are the underlying mechanisms with which systems of interacting electrons may produce a variety of magnetic structures.

2.1 Ferromagnetic Exchange

We consider two electrons in two nearly degenerate single-particle orbitals ϕ_1 and ϕ_2 with energies ϵ_1 and ϵ_2, respectively. We assume that the rest of the spectrum is separated from these states by a large energy gap. In Fig. 2.1 we depict a one-dimensional example for such states. In reality, they might be members of a low-lying angular momentum multiplet of an atom or a molecule. We also assume that we have two electrons which can occupy either or both of these states. The degeneracy of the different configurations is lifted by the Coulomb interactions. We write the electron fields in the ϕ_i

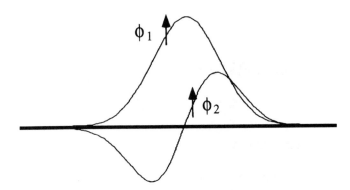

FIGURE 2.1. States of two electrons that couple ferromagnetically.

basis:

$$\psi_s^\dagger(x) = \sum_{i=1}^{2} \phi_i^* c_{is}^\dagger, \qquad s = \uparrow, \downarrow \ . \tag{2.1}$$

The two-body Coulomb interaction in the second quantized form is given by (see (1.17))

$$\mathcal{V} = \frac{1}{2} \int d^3x \, d^3y \ \tilde{v}(\mathbf{x}, \mathbf{y}) \sum_{ss'} \psi_s^\dagger(\mathbf{x}) \psi_{s'}^\dagger(\mathbf{y}) \psi_{s'}(\mathbf{y}) \psi_s(\mathbf{x}). \tag{2.2}$$

We use (2.1) to express (2.2) in the i representation as

$$\mathcal{V} = \sum_{i \neq i'} U_{ii'} \rho_i \rho_{i'} + \sum_{i} U_{ii} \rho_{i\uparrow} \rho_{i\downarrow} + \sum_{ss', i \neq i'} J^F c_{is}^\dagger c_{i's'}^\dagger c_{is'} c_{i's} \ , \tag{2.3}$$

where $n_i = \sum_s \rho_{is}$. The interaction parameters are given by the *direct* integrals:

$$U_{ii'} = \frac{1}{2} \int d^3x \, d^3y \ \tilde{v}(\mathbf{x}, \mathbf{y}) \ |\phi_i(\mathbf{x})|^2 |\phi_{i'}(\mathbf{y})|^2 \tag{2.4}$$

and the *ferromagnetic exchange* constant:

$$J^F = \frac{1}{2} \int d^3x \, d^3y \ \tilde{v}(\mathbf{x}, \mathbf{y}) \ \phi_i(\mathbf{x}) \phi_{i'}^*(\mathbf{x}) \phi_{i'}(\mathbf{y}) \phi_i^*(\mathbf{y}) \ . \tag{2.5}$$

It is easy to see that $U_{ii'}$ is positive and that J is real. It is obvious that for a short-range interaction $\tilde{v} = \delta(\mathbf{x} - \mathbf{y})$, $J^F > 0$. For long-range Coulomb interactions $v^c = e^2/|\mathbf{x} - \mathbf{y}|$, the positivity of J is proved as follows. The eigenvalues of the symmetric operator $v^c(\mathbf{x}, \mathbf{y})$ are given by

$$\int d^3x \ e^{i\mathbf{k}\mathbf{x}} \frac{e^2}{|\mathbf{x}|} = \frac{4\pi e^2}{k^2} > 0. \tag{2.6}$$

Thus, all expectation values of v^c are positive, particularly the expectation value in the state $\Phi(\mathbf{x}) = \phi_{i'}^* \phi_i$, i.e., (2.5) is positive. Thus we have shown that (2.5) is positive for two limiting cases: (i) complete screening, and (ii) no screening at all.

Now, if

$$\epsilon_1 + \epsilon_2 + U_{12} < \min \left[2\epsilon_1 + U_{11}, 2\epsilon_2 + U_{22}\right), \tag{2.7}$$

then the ground state mostly contains two electrons occupying different orbitals, i.e., $n_i \approx 1$ and the low-lying states are given by

$$\{|s_1, s_2\rangle\}, \quad s_i = \uparrow, \downarrow \quad . \tag{2.8}$$

The exchange interaction J^F acts in the space (2.8) as a Heisenberg interaction:

$$J^F \sum_{ss'} c_{is}^\dagger c_{i's'}^\dagger c_{is'} c_{i's} = -2J^F \left(\mathbf{S}_i \cdot \mathbf{S}_{i'} + \frac{1}{4} n_i n_{i'}\right). \tag{2.9}$$

$\mathbf{S}_i^\alpha$, $\alpha = x, y, z$ are the components of spin one half operators (see (A.19)):

$$\mathbf{S}_i \equiv \frac{1}{2} \sum_{ss'} c_{is}^\dagger \vec{\sigma}_{ss'} c_{is'} \quad . \tag{2.10}$$

The exchange integrals (2.5) depend on the spatial overlap between the orbitals. Equations (2.3) and (2.9) explicitly demonstrate the ferromagnetic coupling between electron spins that occupy such orbitals.

A physical argument: By aligning with each other and forming a symmetric spin state, the spins reduce the effect of the Coulomb repulsion. The two electron state is antisymmetric, and therefore their orbital wave function would have to vanish at $\mathbf{x} - \mathbf{y} = 0$, where the Coulomb potential is largest.

This effect appears at first order in perturbation theory and shares the same underlying physics as *Hund's rules* for atoms: Electrons occupy orbitals in an open shell so as to maximize their total spin.[1]

2.2 Antiferromagnetic Exchange

The simplest system that exhibits antiferromagnetic coupling between two electrons is the H_2 molecule, which was first discussed by Heitler and London.[2]

---

[1] The second Hund's rule says that ambiguities in the first rule are settled by maximizing the total orbital angular momentum.

[2] Here we follow the discussion in Mattis' book.

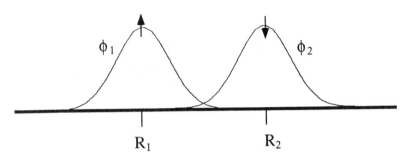

FIGURE 2.2. States of two electrons that couple antiferromagnetically.

Let us freeze the protons at their equilibrium position $\mathbf{R}_i$, and denote the electron coordinates at $\mathbf{x}_i$, for $i = 1, 2$. The full Hamiltonian is

$$H = \sum_{i=1}^{2} H_i^{\infty} + \Delta H \ , \tag{2.11}$$

$$H_i^{\infty} = -\frac{\hbar^2}{2m} \nabla_i^2 - \frac{e^2}{|\mathbf{x}_i - \mathbf{R}_i|^2} \ ,$$

$$\Delta H = -\sum_{i \neq j} \frac{e^2}{|\mathbf{x}_i - \mathbf{R}_j|^2} + \frac{e^2}{|\mathbf{x}_1 - \mathbf{x}_2|^2} + \frac{e^2}{|\mathbf{R}_1 - \mathbf{R}_2|^2}. \tag{2.12}$$

The two atomic wave functions are ϕ_1 and ϕ_2, which are centered on atoms 1 and 2, respectively, as depicted in Fig. 2.2. They obey the atomic Schrödinger equations:

$$H_i^{\infty} \phi_i(\mathbf{x}_i) = e_0 \phi_i(\mathbf{x}_i), \quad i = 1, 2. \tag{2.13}$$

For finite interatomic distance $|\mathbf{R}_1 - \mathbf{R}_2| < \infty$, ϕ_1 and ϕ_2 are not orthogonal and their overlap is

$$\lambda = \int d^3 x \, \phi_1^*(\mathbf{x}) \phi_2(\mathbf{x}). \tag{2.14}$$

We use ϕ_i to construct the orbital wave functions

$$\psi^a = \phi_1(\mathbf{x}_1) \phi_2(\mathbf{x}_2) \ , \tag{2.15}$$

$$\psi^b = \phi_1(\mathbf{x}_2) \phi_2(\mathbf{x}_1), \tag{2.16}$$

with χ as a yet unknown spin wave function. Equation (2.16) describes electrons at different sites. We can use them to construct variational wave functions that avoid the large on-site Coulomb repulsion (later we shall enforce the antisymmetry constraint using the spin part of the wave function),

$$\Psi = c^a \psi^a + c^b \psi^b. \tag{2.17}$$

The variational energy is given by

$$E = \langle \Psi | H | \Psi \rangle / \langle \Psi | \Psi \rangle. \tag{2.18}$$

Minimizing $E[c^a, c^b]$ with respect to the c's, we obtain a matrix equation:

$$\left[\begin{pmatrix} U^d & U^o \\ (U^o)^* & U^d \end{pmatrix} - (E - 2e_0) \begin{pmatrix} 1 & \lambda^2 \\ \lambda^{*2} & 1 \end{pmatrix} \right] \begin{pmatrix} c^a \\ c^b \end{pmatrix} = 0, \tag{2.19}$$

where

$$U^d = \int d^3x_1 \, d^3x_2 \, \Delta H \, |\psi^a|^2 , \tag{2.20}$$

$$U^o = \int d^3x_1 \, d^3x_2 \Delta H \, (\psi^a)^* \psi^b \tag{2.21}$$

are the diagonal and off-diagonal terms. Since ψ^a and ψ^b can be made real (they are made out of hydrogen atom ground states), λ and U^o can also be made real. The solutions of (2.19) are

$$E^\pm = 2e_0 + \frac{U^d \pm U^o}{1 \pm \lambda^2} , \tag{2.22}$$

$$\begin{pmatrix} c^a \\ c^b \end{pmatrix}^\pm = \frac{1}{\sqrt{2}} \begin{pmatrix} 1 \\ \mp 1 \end{pmatrix}. \tag{2.23}$$

Now we impose the antisymmetry condition: the product of Ψ and the spin part of the wave function $\chi(\mathbf{x}_1, \mathbf{x}_2)$ must be antisymmetric with respect to exchange of the electrons coordinates:

$$\begin{aligned} \Psi^\pm &= \frac{1}{\sqrt{2}} (\psi^a \mp \psi^b) \chi^\pm \\ &= \frac{1}{\sqrt{2}} [\phi_1(\mathbf{x}_1) \phi_2(\mathbf{x}_2) \mp \phi_1(\mathbf{x}_2) \phi_2(\mathbf{x}_1)] \chi^\pm , \\ \chi^\pm &= \chi^\uparrow(\mathbf{x}_1) \chi^\downarrow(\mathbf{x}_2) \pm \chi^\uparrow(\mathbf{x}_2) \chi^\downarrow(\mathbf{x}_1) , \end{aligned} \tag{2.24}$$

where $\chi^\pm$ describe antiparallel spins. It is readily verified that $\Psi^\pm$ are linear combinations of two Slater determinants, which can be written as Fock states.[3] Thus, in second quantized notations we can write

$$\Psi^\pm = \frac{1}{\sqrt{2}} \left(c_{1\uparrow}^\dagger c_{2\downarrow}^\dagger \pm c_{1\downarrow}^\dagger c_{2\uparrow}^\dagger \right) |0\rangle. \tag{2.25}$$

---

[3]Strictly speaking, in order to define these as Fock states, ϕ_1 and ϕ_2 must be first orthogonalized.

It is apparent from this form that Ψ^+ and Ψ^- are simply the triplet and singlet eigenstates of the total spin operator

$$\mathbf{S}^{tot} = \sum_{i=1}^{2} \frac{1}{2} \sum_{ss'} c_{is}^\dagger \vec{\sigma}_{ss'} c_{is'} \,. \tag{2.26}$$

The energy splitting between the singlet and the triplet is given by

$$E^+ - E^- = J$$
$$J = 2\frac{U^d\lambda^2 - U^o}{1 - \lambda^4}. \tag{2.27}$$

It can be shown, by an explicit calculation of the parameters U^d and U^o, that J is positive (antiferromagnetic) for all R_{12}. Therefore, the ground state is the singlet, and the higher state is the triplet, and the effective Hamiltonian can be represented by the Heisenberg *antiferromagnet*,

$$\mathcal{H}^{QHM} = J\mathbf{S}_1 \cdot \mathbf{S}_2$$
$$= J\left[\frac{1}{2}(\mathbf{S}^{tot})^2 - \frac{3}{4}\right] = \begin{cases} \frac{1}{4}J & S^{tot} = 1 \\ -\frac{3}{4}J & S^{tot} = 0 \end{cases}. \tag{2.28}$$

Physical argument: The antiparallel spins can take advantage of the hybridization and reduce their kinetic energy by hopping to the second site, while the other electron is there. Parallel spins are restricted by the Pauli principle from this virtual process, which makes the triplet higher in energy than the singlet.

Summary: When two electrons are localized on energetically close orbitals and are interacting repulsively, their magnetic coupling will be:

1. Ferromagnetic: if the orbits are orthogonal but occupy the same region in space. Here the alignment of the spins reduces the *interaction energy.*

2. Antiferromagnetic: if the orbits are not orthogonal but spatially separated. Here the anti-alignment of the spins reduces the *kinetic energy.*

2.3   Superexchange

In the previous section, we have seen that two electrons on nearby hydrogen atoms tend to couple antiferromagnetically due to the interplay between on-site repulsion and their delocalization energy. This effect can be derived in a simple manner using second quantized operators by expanding the energy to second order in the hopping matrix element. Let us consider two *orthogonal* orbitals localized on two atoms labelled by $i = 1, 2$. Tunnelling between the two states is described by a hopping Hamiltonian:

$$\mathcal{H}^t = -t \sum_s \left(c_{1s}^\dagger c_{2s} + c_{2s}^\dagger c_{1s} \right). \tag{2.29}$$

For simplicity, we consider an on-site (Hubbard) interaction:

$$\mathcal{U} = U \sum_i n_{i\uparrow} n_{i\downarrow} . \tag{2.30}$$

For large values of U/t, we choose $\mathcal{U}$ to be our zeroth-order Hamiltonian, whose ground state manifold is fourfold degenerate:

$$\{0\} = |s_1, s_2\rangle , \quad s_i = \uparrow, \downarrow, \quad i = 1, 2. \tag{2.31}$$

The energies of the doubly occupied orbitals are higher by U than the multiplet (2.31). The first-order perturbation theory in $\mathcal{H}^t$ takes us out of the ground state manifold, since it puts one electron on top of the other. Second-order perturbation theory in the subspace (2.31) yields an effective Hamiltonian given by the standard expression[4]

$$\langle a|\mathcal{H}^{(2)}|b\rangle = -\langle a|\mathcal{H}^t \frac{1 - P_0}{\mathcal{U}} \mathcal{H}^t|b\rangle = -\sum_{n \notin \{0\}} \langle a|\mathcal{H}^t|n\rangle \frac{1}{\langle n|\mathcal{U}|n\rangle} \langle n|\mathcal{H}^t|b\rangle, \tag{2.32}$$

where a, b denote states in the subspace $\{0\}$ whose projector is P_0. Thus $|n\rangle$ contain double occupancies. Each term in this sum can be represented by an "exchange path." For example,

$$\begin{array}{cccc} & \mathcal{H}^t & \frac{1}{\mathcal{U}} & \mathcal{H}^t \\ |\uparrow, \downarrow\rangle & \rightarrow & |\uparrow\downarrow, 0\rangle & \rightarrow & |\downarrow, \uparrow\rangle \\ & \rightarrow & |0, \uparrow\downarrow\rangle & \rightarrow & |\downarrow, \uparrow\rangle. \end{array} \tag{2.33}$$

Each of the two paths in (2.33) yields a contribution $-t^2/U$. However, there are paths that are blocked by the Pauli exclusion principle, such as

$$\begin{array}{cc} & \mathcal{H}^t \\ |\uparrow, \uparrow\rangle & \rightarrow \quad 0. \end{array} \tag{2.34}$$

Therefore, the triplet states do not gain second-order energy by virtual double occupation. The operators that connect these initial and final states can be written as products of S=1/2 spin operators $S_1^- S_2^+$. When the paths that describe the matrix elements of $S_2^- S_1^+$ and $S_1^z S_2^z$ are included, we find

---

[4]See Landau and Lifshitz, the bibliography.

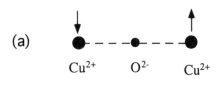

(a)

Cu^{2+} O^{2-} Cu^{2+}

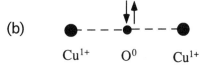

(b)

Cu^{1+} O^0 Cu^{1+}

FIGURE 2.3. Superexchange in copper oxides. (a) A member of the low-energy manifold. (b) A high energy configuration in the superexchange path.

that the effective Hamiltonian in the ground state subspace can be written as an isotropic antiferromagnetic Heisenberg exchange:

$$\mathcal{H}^{(2)} = J\mathbf{S}_1 \cdot \mathbf{S}_2 , \quad J = \frac{4t^2}{U} . \tag{2.35}$$

This discussion is easily generalized to a multi-site system. The antiferromagnetic coupling will be generated between any two spins that are coupled by a hopping term $\mathcal{H}^t$.

The process of virtual double occupation is called *"superexchange."* The analysis of the hydrogen molecule given in the previous section used a variational method to describe the same effect. We distinguish the superexchange from the ferromagnetic direct exchange or Hund's rule given in Section 2.1. The latter is a consequence of *first-order* perturbation in the interactions.[5].

Superexchange couplings are generated in many other systems with unpaired spins. Notable examples with much current interest are the copper-oxide antiferromagnets. The unpaired spins are on Cu^{2+} ions, and between them there is an oxygen "bridge" ion, with average ionization state close to O^{2-}. We illustrate such a triad in Fig. 2.3.

In the absence hopping between the coppers and oxygens, the degenerate ground state manifold on-bond is

$$\{|0\rangle\} = \{|Cu^{2+}s_1, O^{2-}, Cu^{2+}s_2\rangle , \quad s_i = \uparrow, \downarrow . \tag{2.36}$$

The lowest-order correction to the ground state energy that lifts the degeneracy of (2.36) is *fourth order* in the copper–oxygen hopping. Examples

---

[5]In his original paper (see bibliography), Anderson considered several exchange mechanisms. There, superexchange was defined as the coupling that involved an intermediate ion.

of intermediate states in the superexchange paths and their energies are

$$|Cu^{1+}, O^{2-}, Cu^{3+} \uparrow\downarrow\rangle, \quad E = U_d, \tag{2.37}$$

and another, with different energy denominator, is

$$|Cu^{1+}, O \uparrow\downarrow, Cu^{1+}\rangle, \quad E = 2(\epsilon_p - \epsilon_d) + U_p, \tag{2.38}$$

where $\epsilon_{d(p)}$ and $U_{d(p)}$ are the single-particle energy and interaction energy on the copper (oxygen), respectively. As in (2.34), the parallel spin states are blocked from hopping into either of these intermediate states. This lifts the degeneracy between the singlet and the triplet states in (2.36), which is described by a Heisenberg Hamiltonian. The antiferromagnetic coupling is given by the sum over all superexchange paths. It is dominated therefore by the intermediate states with the lowest energy.

2.4   Exercises

1. The fermion representation of spin operators is given in (2.10). Prove the important identity in (2.9):

$$\sum_{ss'} c_{1s}^{\dagger} c_{2s'}^{\dagger} c_{1s'} c_{2s} = -\frac{1}{2}\{n_1 n_2 + 4\mathbf{S}_1 \cdot \mathbf{S}_2\}, \tag{2.39}$$

   where $n_i = \sum_s c_{is}^{\dagger} c_{is}$, $i = 1, 2$.

2. Prove that $\mathbf{S}_i$ obey the angular momentum commutation relations. Find a simple expression for $\mathbf{S}_i^2$ as a function of n_i.

Bibliography

The hydrogen molecule analysis follows Mattis' book quite closely:

• D.C. Mattis, *The Theory Of Magnetism I* (Springer-Verlag, 1988).

The theory of superexchange was proposed by Kramers, and developed by Anderson in his seminal paper,

• P.W. Anderson, Phys. Rev. **79**, 350 (1950).

Superexchange coupling for copper oxide superconductors has been worked out by, e.g.,

• F.C. Zhang and T.M. Rice, Phys. Rev. B **37**, 3759 (1988).

For the basics of perturbation theory see

• L.D. Landau and E.M. Lifshitz, *Quantum Mechanics* (Pergamon Press, 1977), Chapter VI.

3

The Hubbard Model and Its Descendants

Many theorists have devoted a considerable part of their careers to the Hubbard model. Nevertheless, it remains a source of much fascination and bewilderment. Perhaps it is because the Hubbard model is the simplest many-particle model one can write down, which cannot be reduced to a single-particle theory.

The ground state is known to be complicated (i.e., a superposition of many Fock states). In most cases its analytic form is unknown, except in one dimension (see the bibliography). There the ground state correlations and excitations have been understood in several ways. The methods used for one dimension are innovative and diverse: the Bethe ansatz solution, bosonization, field theoretical methods, the Luttinger and Tomonaga models, and the perturbative renormalization group. Unfortunately, the methods listed above are specialized to one dimension and their application to two and three dimensions has yet to produce conclusive results.

Nevertheless, in two and three dimensions, progress has been made in several regimes of parameter space using combinations of theorems, controlled approximation schemes, and numerical results extrapolated to large lattices. For further background, see the bibliography for a recent book on the Hubbard model.

We shall see in Section 3.1 that some of the truncations in the Hubbard model are justified in the "atomic limit." It incorporates the *short-range* part of the Coulomb interactions, while avoiding the high complexity (such as screening effects) of the long-range Coulomb force.

However, the Hubbard model ignores terms that are not obviously small in the microscopic Hamiltonian. It should therefore be regarded as an *effective* Hamiltonian which includes only the most relevant couplings at low energies. Its parameters could, in principle, be determined by integrating out the high-energy modes in a renormalization group scheme. Unfortunately, this is a hard task which has not yet been carried out quantitatively for realistic band structures.

In the strongly interacting (large-U) limit at half filling, the Hubbard Hamiltonian is reduced to a purely spin model as we shall see in Section 3.2. In Section 3.3 we discuss the attractive (negative-U) Hubbard model. This model is used for understanding superconductivity and charge density wave ordering. We prove that the negative-U model maps into a positive-U model at half filling. Thus, the negative-U model is translated into a problem of quantum magnetism.

3.1   Truncating the Interactions

In the following, we shall attempt to partially justify the truncation of terms that are missing in the Hubbard model. It would be correct, but of little use, to start with the bare ion–electron potential in H^0 and the long-range Coulomb interaction $v(r) = e^2/r$ for the interaction terms. The collective screening of the core and valence electrons is large. As discussed in Section 1.1, much of the screening effect is to renormalize the effective electron–ion potential. The interactions are thus taken to be

$$\tilde{\mathcal{V}}^{el-el} = \frac{1}{2} \int d^3x \, d^3y \, \tilde{v}(\mathbf{x}, \mathbf{y}) \sum_{ss'} \psi_s^\dagger(\mathbf{x}) \psi_{s'}^\dagger(\mathbf{y}) \psi_{s'}(\mathbf{y}) \psi_s(\mathbf{x}). \tag{3.1}$$

Here spin–orbit interactions (relativistic corrections) were ignored for simplicity. We should remember, however, that these interactions can be important for obtaining the correct magnetic moment crystal field splittings and anisotropic exchanges for d and f shells.

We use the band structure energies and Bloch wave functions given by the self-consistent single-particle Hamiltonian (1.7):

$$\tilde{\mathcal{H}}^0 \rightarrow \{\epsilon_{\alpha\mathbf{k}}, \phi_{\alpha\mathbf{k}}\}, \tag{3.2}$$

where α is the band index. The Wannier states derived from $\phi_{\alpha\mathbf{k}}$ are

$$\phi_{\alpha i}(\mathbf{x}) \equiv \frac{1}{\sqrt{N}} \sum_{\mathbf{k}} e^{-i\mathbf{k}\mathbf{x}} \, \phi_{\alpha\mathbf{k}}(\mathbf{x}), \tag{3.3}$$

where N is the number of lattice sites. Wannier states are a single-particle basis labelled by the lattice site index i and the band index α. The Wannier operators are

$$c_{\alpha i s}^\dagger \equiv \int d^3x \, \phi_{\alpha i}(\mathbf{x}) \psi_s^\dagger(\mathbf{x}). \tag{3.4}$$

Since the transformation (3.4) is unitary, the sets $\{c_{\alpha i s}\}, \{c_{\alpha i s}^\dagger\}$ obey canonical anticommutation relations (see (A.10)). Inverting (3.4) we have

$$\psi_s^\dagger(\mathbf{x}) = \sum_{i\alpha} \phi_{\alpha i}^* c_{\alpha i s}^\dagger. \tag{3.5}$$

The field operators in the interacting Hamiltonian (1.16 and 1.17) are replaced by Wannier operators to yield

$$\mathcal{H} = -\sum t_{\alpha i j} c_{\alpha i s}^\dagger c_{\alpha j s} + \sum U_{ijkl}^{\alpha\beta\gamma\delta} c_{\alpha i s}^\dagger c_{\beta j s'}^\dagger c_{\gamma k s'} c_{\delta l s} , \tag{3.6}$$

where the summations are over all repeated indices. t_{ij} are the *hopping* matrix elements given by

$$t_{\alpha ij} = -\langle \phi_{\alpha is} | \tilde{H}^0 | \phi_{\alpha is} \rangle = \frac{1}{N} \sum_{\mathbf{k}} e^{-i(\mathbf{x}_j - \mathbf{x}_i)\mathbf{k}} \epsilon_{\mathbf{k}}^\alpha , \qquad (3.7)$$

which by hermiticity of the Hamiltonian obey $t_{ij} = t_{ji}^*$. In the absence of external gauge fields, t_{ij} can be chosen to be real. The *interaction parameters* are given by

$$U_{ijkl}^{\alpha\beta\gamma\delta} = \frac{1}{2} \int d^3x \, d^3y \; \tilde{v}(\mathbf{x}, \mathbf{y}) \phi_{\alpha i}^*(\mathbf{x}) \phi_{\beta j}^*(\mathbf{y}) \phi_{\gamma k}(\mathbf{y}) \phi_{\delta l}(\mathbf{x}). \qquad (3.8)$$

An optimal choice of Wannier states (through the choice of $\tilde{\mathcal{H}}^0$) would minimize the range and magnitude of U_{ijkl}. From this point onward, many terms in (3.6) will be omitted.

When the Fermi surface lies within a single conduction band, $\alpha = 1$, it may be justified to ignore matrix elements that couple to other bands if they are well separated from the Fermi energy. This truncation leads to the *one-band Hubbard model.*

For f-electron metals (rare earth compounds), the single-band model is not justified since the interaction parameters are larger than the interband splittings. For those systems, there are charge fluctuations between the deeply localized f levels and the delocalized s and p bands. Such models are given by the Anderson and Kondo Hamiltonians, which have been used to describe mixed valence and heavy Fermion systems (see bibliography).

There are two classes of interactions that are omitted: the direct terms and the exchange terms.

The direct terms involve integrals over *square moduli* of Wannier functions (i.e., $|\phi_i|^2$ in (3.8)). They are

$$\mathcal{V} = \sum_{i \neq j} V_{ij} n_i n_j. \qquad (3.9)$$

The intersite interactions $V_{ij} = U_{ijji}$ couple *density fluctuations* at different sites. V_{ij} are not necessarily smaller than t_{ij}. When the interactions are poorly screened they do not decrease rapidly with distance. Thus, by considerations of magnitude, we are not justified in neglecting $\mathcal{V}$. On the other hand, we note that $\mathcal{V}$ couples charge fluctuations rather than spin fluctuations. In the vicinity of *magnetic* phase transitions, the contribution of such terms to the free energy is not singular. They are expected to be relevant near *charge density* instabilities.

The exchange terms include intersite magnetic couplings. For example, the two-site terms can be written as

$$\mathcal{J} = \sum_{i \neq j} U_{ijij} c_{is}^\dagger c_{js'}^\dagger c_{is'} c_{js} = -\frac{1}{2} \sum_{i \neq j} J_{ij}^F \left(\mathbf{S}_i \mathbf{S}_j + \frac{1}{4} n_i n_j \right), \qquad (3.10)$$

where $J_{ij}^F = U_{ijij}$. This ferromagnetic exchange term has been previously found in (2.9) for two electrons in two orbitals. As argued by Hirsch (see bibliography), these terms can contribute to ferromagnetic tendencies in transition metals. Nevertheless, they are customarily ignored in the Hubbard model, and the reasoning is as follows. The exchange integrals involve one or two off-diagonal products of Wannier functions (i.e., $\phi_i^* \phi_j.$). The intersite ferromagnetic exchanges such as (3.10) are suppressed in the *atomic limit*, where the Wannier orbitals are approximately superpositions of atomic orbitals[1]:

$$\phi_{\mathbf{k}}(\mathbf{x}) \simeq \sum_i e^{i\mathbf{k}\mathbf{x}_i} \phi(\mathbf{x} - \mathbf{x}_i). \tag{3.11}$$

This cannot be an equality since atomic wave functions in the lattice are not orthogonal. There overlaps are of order

$$\int d^3x \, |\phi_i(\mathbf{x})^* \phi_j(\mathbf{x})| \simeq e^{-|\mathbf{x}_i - \mathbf{x}_j|/l_a} , \tag{3.12}$$

where l_a is the average radius of the atomic orbital.

Thus, the parameters t_{ij} and J_{ij}, which involve integrals over overlap factors, are also small. The atomic limit is consistent with two additional limits of the Hubbard model:

1. *The tight binding limit*, where one retains a minimal set of short-range bonds $\{t_{ij}\}$ on the lattice.

2. *The large-U limit.* The on-site Hubbard interaction increases as the atomic wave functions become more localized. Thus, we expect

$$U \simeq \frac{e^2}{l_a} >> t_{ij} . \tag{3.13}$$

In conclusion, the minimal Hubbard model for a band structure in the atomic limit is given by the tight binding hopping parameters t_{ij}, the onsite interaction $U = 2U_{iiii}$, and the number of electrons N_e:

$$\mathcal{H} = -\sum_{ijs} t_{ij} c_{is}^\dagger c_{js} + U \sum_i n_{i\uparrow} n_{i\downarrow}. \tag{3.14}$$

The simplicity of (3.14) is highly deceiving. The interaction term gives rise to highly correlated ground states. In the following chapters, we shall study different types of magnetic correlations that are produced in this model.

---

[1] In this limit, the suppressed band index corresponds to an atomic energy level.

3.2   At Large U: The $t-J$ Model

Here we derive the effective Hamiltonian, which governs the low-energy excitations of the Hubbard model in the large U/t regime. We introduce the concept of an effective Hamiltonian for the low-energy excitations and expand it in powers of t/U. At half filling (one electron per site), for $U/t >> 1$ there are primarily spin excitations at low energies. These will be shown to be governed by the antiferromagnetic Heisenberg model of spin one half. At other fillings, there are low-energy spin and charge excitations which are governed by the so-called $t-J$ *model.*

We start by choosing the zeroth-order Hamiltonian to be

$$\mathcal{U} = U \sum_i n_{i\uparrow} n_{i\downarrow}. \tag{3.15}$$

The eigenstates of $\mathcal{U}$ are Fock states in the Wannier representation. $\mathcal{U}$ divides the Fock space into two subspaces:

$$S = \left[|n_{1\uparrow}, n_{1\downarrow}, n_{2\uparrow} \ldots\rangle \quad : \quad \forall i, \ n_{i\uparrow} + n_{i\uparrow} \leq 1 \right], \tag{3.16}$$

$$D = \left[|n_{1\uparrow}, n_{1\downarrow}, n_{2\uparrow} \ldots\rangle \quad : \quad \exists i, \ n_{i\uparrow} + n_{i\uparrow} = 2 \right]. \tag{3.17}$$

D contains at least one doubly occupied site, and S are all configurations with either one or zero electrons per site.

The hopping term

$$\mathcal{T} = -\sum_{ijs} t_{ij} c_{is}^\dagger c_{js} \tag{3.18}$$

is considered as a perturbation. $\mathcal{T}$ couples S to D by hopping an electron into, or out of, a doubly occupied site. $\mathcal{U}$ is diagonal and $\mathcal{T}$ lifts the enormous degeneracy in the two subspaces. $\mathcal{H}$ is partitioned as

$$\mathcal{H} = \begin{pmatrix} P_s(\mathcal{T}+\mathcal{U})P_s & P_s \mathcal{T} P_d \\ P_d \mathcal{T} P_s & P_d(\mathcal{T}+\mathcal{U})P_d \end{pmatrix}, \tag{3.19}$$

where $P_{s,d}$ are projection operators onto the subspaces S and D, respectively.

The resolvent operator is

$$\mathcal{G}(E) = (E - \mathcal{H})^{-1}. \tag{3.20}$$

The following matrix identity is useful:

$$\left[\begin{pmatrix} A & B \\ C & D \end{pmatrix}^{-1} \right]_{ss} = (A - BD^{-1}C)^{-1}, \tag{3.21}$$

where $(\;)_{ss}$ denotes singly occupied subspace in the upper left quadrant. The projection of $\mathcal{G}$ into the subspace $\mathcal{S}$ is

$$P_s \mathcal{G}(E) P_s = P_s [E - H]^{-1} P_s = [E - \mathcal{H}^{eff}(E)]^{-1}. \qquad (3.22)$$

Using (3.21) and that $P_s \mathcal{U} = \mathcal{U} P_s = 0$, the effective Hamiltonian is given explicitly by

$$\begin{aligned} \mathcal{H}^{eff} &= P_s \mathcal{T} P_s \\ &\quad + P_s \mathcal{T} \{ P_d [E - (\mathcal{U} + \mathcal{T})] P_d \}^{-1} \mathcal{T} P_s. \end{aligned} \qquad (3.23)$$

The eigenenergies of $\mathcal{H}$ (which correspond to states with nonzero weights in $\mathcal{S}$) are given by the zeros of the characteristic polynomial

$$\det \left| E_n - \mathcal{H}^{eff}(E_n) \right| = 0. \qquad (3.24)$$

Note that E_n are not eigenvalues of $\mathcal{H}^{eff}$, since $\mathcal{H}^{eff}$ depends parametrically on E. If we ignore (for the moment) problems arising from a large number of sites, it is possible to expand $P_d (E - \mathcal{H})^{-1} P_d$, to zeroth order in E/U and to second order in t/U. This allows us to replace (3.24) by an eigenvalue equation of an *energy independent Hamiltonian*:

$$\mathcal{H}^{eff} \;\rightarrow\; \mathcal{H}^{t-J} \left[1 + \mathcal{O}(E/U) + \mathcal{O}(t/U) \right] \;,$$

$$\mathcal{H}^{t-J} = P_s \left[\mathcal{T} - \frac{1}{U} \sum_{ijkss'} t_{ij} t_{jk} c_{is}^{\dagger} c_{js} n_{j\uparrow} n_{j\downarrow} c_{js'}^{\dagger} c_{ks'} \right] P_s. \qquad (3.25)$$

H^{t-J} is commonly called *the t−J model*.

Let us pause for a minute to recognize that for a large number of lattice sites $\mathcal{N} \rightarrow \infty$. When the number of electrons scales as $\mathcal{N}$, the ground state energy E_0 is *extensive*, i.e., it involves a sum over zero-point energies which scales as $\mathcal{N}$. Hence it is not sensible to expand $\mathcal{H}^{eff}(E)$ about $E = 0$. However, it is still possible to derive an effective Hamiltonian for low-lying excitations by shifting the zero of energy to

$$E^0 \;\rightarrow\; E_0' = E_0^d - U. \qquad (3.26)$$

E_0^d, which is also extensive, is the lowest energy of the Hubbard model in the doubly occupied sector. We do not need to actually calculate E_0^d if we forego the knowledge of the Hubbard model's ground state energy. We therefore restrict our use of $\mathcal{H}^{t-J}$ to describe low-lying elementary excitations and wave functions. Now, we repeat the procedure described above and expand the energy for small

$$|E - E_0'| \ll U. \qquad (3.27)$$

This allows us to study the low-frequency and temperature response using an energy independent effective Hamiltonian.[2]

A more transparent form for the $t - J$ model is obtained by rearranging the fermion operators:

$$\mathcal{H}^{t-J} = P_s \left(\mathcal{T} + \hat{\mathcal{H}}^{QHM} + \mathcal{J}' \right) P_s, \tag{3.28}$$

$$\mathcal{T} = -\sum_{ijs} t_{ij} c_{is}^\dagger c_{js}, \tag{3.29}$$

$$\mathcal{H}^{QHM} = \frac{1}{2} \sum_{ij} J_{ij} \left(\mathbf{S}_i \cdot \mathbf{S}_j - \frac{n_i n_j}{4} \right), \tag{3.30}$$

$$\mathcal{J}' = -\frac{1}{2U} \sum_{ijk}^{i \neq k} t_{ij} t_{jk} \left[\sum_s (c_{is}^\dagger c_{ks} n_j) - c_i^\dagger \vec{\sigma} c_k \cdot c_j^\dagger \vec{\sigma} c_j \right], \tag{3.31}$$

where the spin operators $\mathbf{S}_i$ have been defined in (2.10), and the *superexchange* coupling constants are

$$J_{ij} = 4t_{ij}^2/U. \tag{3.32}$$

At half filling we have $n_i = 1$. In that limit, P_s annihilates $\mathcal{T}$ and $\mathcal{J}'$. (There can be no low-energy hopping processes if every site has exactly one electron.) This is a *Mott insulator* phase. The only terms that survive the projection P_s are the magnetic interactions of $\mathcal{H}^{QHM}$, the *quantum Heisenberg model*. Thus we find antiferromagnetic interactions in the Hubbard model which are important (at least for large U/t) near half filling. We shall discuss the Heisenberg Hamiltonian extensively in this book.

The effects of the additional electrons or holes (as measured from the half-filled limit) on the quantum antiferromagnet are a current subject of intense investigations, particularly in the context of high-temperature superconductivity in copper oxides.

One of the effects of doping is believed to be the reduction of antiferromagnetic correlations. Loosely speaking, the holes prefer a ferromagnetic spin environment where they can lower their kinetic energy by being more "mobile." This effect is most prominent in the infinite-U limit, as shown by Nagaoka's theorem in Section 4.2.

In the literature, the $\mathcal{J}'$ terms are frequently omitted. It has been assumed that their effects are small in comparison to the t hoppings. However, for two-sublattices models with nearest neighbor t hoppings, vacancies (or holes) can move under $\mathcal{T}$ between different sublattices, thereby disturbing the antiferromagnetic correlations. This is loosely illustrated in Fig. 3.1. The $\mathcal{J}'$ hoppings do not "mess up" the spins, since they move holes on the

---

[2]The shift in E_0 is the difference between Brillouin–Wigner and Rayleigh–Schrödinger perturbation theories, see, e.g., Problem 3.6 in J.W. Negele and H. Orland, *Quantum Many Particle Systems* (Addison-Wesley, 1988).

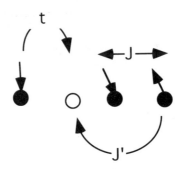

FIGURE 3.1. Interactions between a hole and spins in the $t-J$ model.

same sublattice.

Before leaving the $t-J$ model we write it once more: this time in its normal-ordered form

$$\mathcal{H}^{t-J} = P_s \left[\mathcal{T} - \frac{1}{U} \sum_{ijk} t_{ij} t_{jk} (c_{i\uparrow}^\dagger c_{j\downarrow}^\dagger - c_{i\downarrow}^\dagger c_{j\uparrow}^\dagger)(c_{j\downarrow} c_{k\uparrow} - c_{j\uparrow} c_{k\downarrow}) \right] P_s.$$

(3.33)

In Chapter 19, we shall treat the $t-J$ model using spin-hole coherent states. After defining its large spin generalization, we shall derive the semiclassical theory for a dilute density of holes.

3.3   The Negative-U Model

The negative-U Hubbard model describes local *attractive* interactions between electrons. As in the repulsive case, one should think of this model as an effective model with renormalized parameters which applies at low frequencies and temperatures. When electrons polarize a collective degree of freedom, they can obtain a negative pair-binding energy $-U$ by sharing this polarization.

There have been several polarization mechanisms proposed for producing attraction between electrons. To list a few: lattice deformations (phonons), collective charge oscillations (plasmons), or spin fluctuations (paramagnons). In the negative-U interaction, the timescale for the polarization mechanism (e.g., the Debye frequency for phonons) is considered to be *instantaneous* relative to the hopping time.[3]

---

[3]This is the opposite regime to Migdal–Eliashberg's approximation for superconductivity, see Schrieffer's book.

We discuss a Hamiltonian of the form

$$\mathcal{H}^{-U} = -\sum_{ijs} t_{ij} c_{is}^\dagger c_{js} - \frac{U}{2} \sum_i (n_i - 1)^2$$
$$+ \frac{1}{2} \sum_{ij} V_{ij}(n_i - 1)(n_j - 1) - \mu \sum_i n_i. \qquad (3.34)$$

The negative-U term favors local pairs of up and down spins on the same site. However, the hopping term $t_{ij} c_{is}^\dagger c_{js}$ competes with this tendency since it delocalizes the electrons and unbinds the pairs. The intersite interactions V_{ij} repel electrons at different sites. V_{ij} are important in selecting the ground state especially near half filling.

Our first step is to transform (3.34) to a positive-U model by a *particle-hole transformation only on the down-spin electrons*:

$$c_{i\uparrow} \rightarrow \tilde{c}_{i\uparrow} ,$$
$$c_{i\downarrow} \rightarrow \tilde{c}_{i\downarrow}^\dagger. \qquad (3.35)$$

This is a canonical Bogoliubov transformation (see Section A.1) since the "pseudo-electrons" $\tilde{c}$ obey Fermi anticommutation relations. Using $\tilde{n}_{is}^2 = \tilde{n}_{is}$, (3.34) readily transforms to a *positive-U* Hamiltonian

$$\mathcal{H}^{-U} \rightarrow \tilde{\mathcal{H}}^{+U} = -\sum_{ij} t_{ij}(\tilde{c}_{i\uparrow}^\dagger \tilde{c}_{j\uparrow} - \tilde{c}_{i\downarrow}^\dagger \tilde{c}_{j\downarrow}) + \frac{U}{2} \sum_i (\tilde{n}_i - 1)^2$$
$$+ \frac{1}{2} \sum_{ij} J_{ij}^a \tilde{S}_i^z \tilde{S}_j^z - h \sum_i \tilde{S}_i^z - \mathcal{N} \Delta E, \qquad (3.36)$$

where the *Ising anisotropy* is

$$J_{ij}^a = 4 V_{ij}, \qquad (3.37)$$

the magnetic field is

$$h = 2\mu , \qquad (3.38)$$

and the energy shift is

$$\Delta E = \mu + \frac{1}{2} U. \qquad (3.39)$$

We now define a *bipartite lattice* for the hopping parameters t_{ij}.

Definition 3.1 *A bipartite lattice can be separated into two disjoint sublattices A and B, where t_{ij} connect only $i \in A$ to $j \in B$ or $i \in B$ to $j \in A$.*

Common examples of bipartite lattices are the square and cubic lattices. Nonbipartite parameters are, e.g., the bonds of the triangular and the face centered cubic (FCC) lattices.

Notice that the interaction term in (3.36) is positive and that the hopping matrix elements have spin-dependent signs. This is physically important for nonbipartite lattices where the signs cannot be eliminated by a simple gauge transformation.

3.3.1 THE PSEUDO-SPIN MODEL AND SUPERCONDUCTIVITY

Transformation (3.35) maps the *charge operators* of the negative-U model into *pseudo-spin operators* of the positive-U counterpart. In particular, the local charge fluctuation maps into a pseudo-spin in the z direction,

$$\frac{1}{2}(n_i - 1) \quad \Leftrightarrow \quad \tilde{S}_i^z. \tag{3.40}$$

Thus, a deviation from half filling in the negative-U model translates to a uniform magnetization in the S^z direction. Similarly, a charge density wave in the negative-U model corresponds to a z-spin density wave in the positive-U model.

The pseudo-spin components $\tilde{S}^x, \tilde{S}^y$ correspond to the *pairing operators*

$$\frac{1}{2}(c_{i\uparrow}^\dagger c_{i\downarrow}^\dagger + c_{i\downarrow} c_{i\uparrow}) \quad \Leftrightarrow \quad \tilde{S}_i^x \ ,$$
$$\frac{1}{2i}(c_{i\uparrow}^\dagger c_{i\downarrow}^\dagger - c_{i\downarrow} c_{i\uparrow}) \quad \Leftrightarrow \quad \tilde{S}_i^y. \tag{3.41}$$

Thus, ordering of the pseudo-spins in the x–y plane represents *superconductivity* in the negative-U model,

$$\Psi(\mathbf{x}_i) = \Delta(\mathbf{x}_i) e^{i\phi(\mathbf{x}_i)} \Leftrightarrow \langle \tilde{S}_i^+ \rangle. \tag{3.42}$$

The magnitude of the ordered moments is the BCS order parameter Δ, and their angle in the x–y plane is the superconducting phase ϕ (see Schrieffer's book, bibliography). Ψ is coupled to an external electromagnetic gauge field with a charge of $2e$. The free-energy expansion of a slowly varying $\Psi(\mathbf{x})$ yields the Ginzburg–Landau free-energy functional. The Ginzburg–Landau theory produces much of the macroscopic phenomenology of superconductivity (e.g., flux quantization, persistent currents, etc.).

Note that the chemical potential of $\tilde{\mathcal{H}}^{+U}$ is zero. Now we prove an important theorem.

Theorem 3.2 For any electron filling of $\mathcal{H}^{-U}$, the positive-U model $\tilde{\mathcal{H}}^{+U}$ is exactly at half filling.

The proof is very simple. Consider the transformation

$$\tilde{c}_{is} \to \tilde{c}_{i-s}^\dagger \ , \quad s = \uparrow, \downarrow \ . \tag{3.43}$$

It can easily be verified that $\tilde{\mathcal{H}}^{+U}$ is invariant under this transformation:

$$\mathcal{H}^{+U}[\tilde{c}'] = \mathcal{H}^{+U}[\tilde{c}], \qquad (3.44)$$

while

$$\tilde{n}_i \rightarrow 2 - \tilde{n}'_i. \qquad (3.45)$$

The number of electrons is given by the thermodynamic average

$$\langle \tilde{n}_i \rangle = \frac{1}{Z} \text{Tr}_{\tilde{c}} \left(e^{-\beta \mathcal{H}^{+U}[\tilde{c}]} \tilde{n}_i \right) = \frac{1}{Z} \text{Tr}_{\tilde{c}'} \left[e^{-\beta \mathcal{H}^{+U}[\tilde{c}']} (2 - \tilde{n}'_i) \right], \qquad (3.46)$$

which by the invariance of Tr under (3.43) (since it is a canonical transformation) implies that[4]

$$\mathcal{N}^{-1} \langle \sum_i \tilde{n}_i \rangle = \mathcal{N}^{-1} \langle \sum_i (2 - \tilde{n}_i) \rangle = 1, \qquad (3.47)$$

Q.E.D.

Theorem 3.2 is important since the half-filled limit of the positive-U Hubbard model can be reduced to a *pure spin problem* at large U. Following the derivations of Section 3.2, the effective pseudo-spin Hamiltonian is

$$\tilde{\mathcal{H}}^{+U} \rightarrow \tilde{\mathcal{H}}^{-x-xz} + \mathcal{O}(t^2/U) \ ,$$
$$\tilde{\mathcal{H}}^{-x-xz} = \frac{1}{2} \sum_{ij} \left[\left(J_{ij} + J^a_{ij} \right) \tilde{S}^z_i \tilde{S}^z_j - J_{ij} \left(\tilde{S}^x_i \tilde{S}^x_j + \tilde{S}^y_i \tilde{S}^y_j \right) \right] - h \sum_i \tilde{S}^z_i,$$
$$(3.48)$$

where the superexchange coupling is again $J_{ij} = 4t^2_{ij}/U$. We have dropped the constant energy shift. $\tilde{\mathcal{H}}^{-x-xz}$ is an anisotropic Heisenberg model. It has ferromagnetic couplings for the x,y spin components, and antiferromagnetic couplings between the z components.

The Hamiltonian (3.48) is a model of *quantum magnetism* that can be treated by many of the methods described in this book. In Chapter 11, we learn that the semiclassical approximation is suitable for broken symmetry phases in two and three dimensions. The classical ground state is given by minimizing the classical Hamiltonian $\tilde{\mathcal{H}}^{-x-xz}[\tilde{\mathbf{S}}]$, where $\tilde{\mathbf{S}}$ are vectors of magnitude S. For the nearest neighbor model on nonbipartite lattices, such as the triangular or face centered cubic (FCC) lattices, the z spin couplings are frustrated (see the bottom bond of the left triangle in Fig. 3.2). At the same time, the ferromagnetic x–y couplings can be fully satisfied, as seen in the right triangle. As a result, the classical approximation yields x–y pseudo-spin ordering, i.e., superconductivity at low temperatures. As we

---

[4]Here we avoid subtleties of local spontaneous symmetry breaking, i.e. phase separation, by defining our averages on finite lattices.

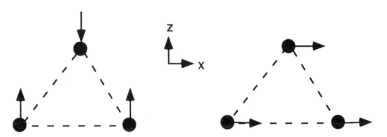

FIGURE 3.2. Frustrated z-spin couplings (left triangle) and satisfied x–y couplings (right triangle) of the pseudo-spin Hamiltonian $\tilde{\mathcal{H}}^{-x-xz}$.

shall see, large quantum and thermal fluctuations can disorder magnetic order especially in low dimensions. Hence, superconductivity is subject to the same disordering mechanisms.

For bipartite lattices,[5] a rotation of π about the z axis of all spins on sublattice B will bring $\tilde{\mathcal{H}}^{-x-xz}$ to the Ising–Heisenberg form

$$\tilde{\mathcal{H}}^{-x-xz} \to \tilde{\mathcal{H}}^{xxz} = \frac{1}{2}\sum_{ij}\left(J_{ij}\tilde{\mathbf{S}}_i \cdot \tilde{\mathbf{S}}_j + J_{ij}^a \tilde{S}_i^z \tilde{S}_j^z\right) - h\sum_i \tilde{S}_i^z. \tag{3.49}$$

The ground state of $\tilde{\mathcal{H}}^{xxz}[\tilde{\mathbf{S}}]$ is antiferromagnetic: the pseudo-spins point in opposite directions on the two sublattices. For $J^a = 0$ and $h = 0$, the classical ground state of (3.49) is $O(3)$-degenerate, as the order parameter can point anywhere on the sphere. This degeneracy is reduced to $O(2)$ for $h > 0$, since a uniform magnetic field in the z direction prefers the spins to lie mostly in the x–y plane, where their canting allows a gain in energy which is linear in h; see Fig. 3.3.

However, at finite Ising anisotropy $J^a > 0$, the spins prefer to align in the z direction even in the presence of a small magnetic field. The competition between the magnetic field and the Ising anisotropy determines the direction of the ordered moments. The transition between the Ising and the x–y ordering may be first or second order in the field h. This transition was well studied theoretically and experimentally in real magnetic systems and applied to theories of superconductivity in bismuthate superconductors and to superfluidity in helium-4 (see bibliography). In the classical mean field approximation, the order of the transition depends on the sign

---

[5]We use definition 3.1 with $t_{ij} \to J_{ij}$.

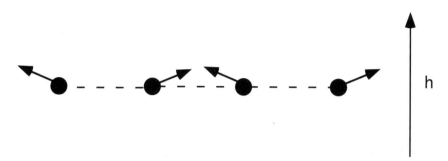

FIGURE 3.3. Spin-flop state of an antiferromagnet in a uniform magnetic field.

of the next nearest neighbor interaction. A first-order transition is called a *spin-flop*, and implies a bicritical point in the field-temperature phase diagram. A second-order transition implies an intermediate mixed phase, where the order parameter has both z and x-y components (i.e., coexisting charge density wave and superconductivity). For helium, the mixed phase is called *"supersolid,"* which to date has not been experimentally discovered.

The effective Hamiltonian (3.48) was derived here for large $|U| >> t$. It describes tightly bound pairs of electrons on a scale of one lattice constant. For weak coupling $|U| < t$, it is better to use perturbation theory in the interaction strength, or variational magnetic states of the kind described in Section 4.1. Both weak and strong coupling are qualitatively similar when there is a gap for charge excitations. At weak coupling, such a gap may arise if large portions of the Fermi surface are parallel (nesting). The pseudo-spin model $\tilde{\mathcal{H}}^{-x-xz}$ can then be used as an effective model, coarse grained over the size of a Cooper pair. That is to say, the "lattice constant" should be replaced by the superconducting correlation length.

3.4 Exercises

1. Prove the identity that was used for deriving the Heisenberg Hamiltonian (3.30):

$$P_s \left[\sum_{ss'} c_{is}^\dagger c_{js} n_{j\uparrow} n_{j\downarrow} c_{js'}^\dagger c_{is'} \right] P_s = -2P_s \left[\mathbf{S}_i \cdot \mathbf{S}_j - \frac{n_i n_j}{4} \right] P_s. \quad (3.50)$$

2. Show, by rearranging the fermion operators, that the $t-J$ model of (3.33) is equal to that of (3.25).

3. Find all the eigenenergies and eigenstates of the two-site Hubbard model (3.14) ($i = 1, 2$), for N_e=1, 2, and 3 electrons. For N_e=2, what is the singlet–triplet splitting? This is the effective spin superexchange constant.

Hint: At each filling N_e, diagonalize the Hamiltonian as a matrix in the relevant Fock states. What is the size of this matrix for a 4×4 lattice?

4. Prove that the Hubbard Hamiltonian (3.14) is rotationally invariant by showing that it commutes with the total spin $\mathbf{S}_{tot} = \sum_i \mathbf{S}_i$. What quantum numbers classify its eigenstates on a one-dimensional ring with $\mathcal{N}$ sites?

Bibliography

A recent comprehensive reference book on the Hubbard model that includes many important papers on the subject is

- *The Hubbard Model*, edited by A. Montorosi (World Scientific, 1992).

For exact solutions in one dimension, see

- *The Many Body Problem — An Encyclopedia of Exactly Solved Models in One Dimension*, edited by D.C. Mattis (World Scientific, 1993).

For field theoretical and renormalization group techniques in one dimension, see

- V.J. Emery *The Many Body Problem In One Dimension*, in *"Correlated Electron Systems"*, edited by V.J. Emery (World Scientific, 1993).
- H.J. Schultz, *Interacting Fermions in One Dimension*, same volume.

A basic textbook on superconductivity is

- J.R. Schrieffer, *Theory of Superconductivity* (Benjamin-Cummings, 1983).

The negative-U model was proposed for bismuthate superconductors by

- T.M. Rice and L. Sneddon, Phys. Rev. Lett. **47**, 689 (1981);
- A.S. Alexandrov, J. Ranninger, and S. Robaszkiewicz, Phys. Rev. B **33**, 4526 (1986);
- A. Aharony and A. Auerbach, Phys. Rev. Lett. **70**, 1874 (1993).

An analysis of the spin-flop transition in the magnetism literature is given in

- *Elements of Theoretical Magnetism*, edited by S. Krupicka and J. Sternberk (Iliffe), Section 7.4;
- M.E. Fisher and D.R. Nelson, Phys. Rev. Lett. **32**, 1350 (1974).

In the context of superfluid helium, the study of the pseudo-spin model and the possibility of a supersolid has a long history:

- T. Matsubara and H. Matsuda, Prog. Theor. Phys. **16**, 569 (1956);

- H. Matsuda and T. Tsuneto, Suppl. Prog. Theor. Phys. **46**, 411 (1970);

- K.S. Liu and M.E. Fisher, J. Low Temp. Phys. **10**, 655 (1973); **17**, 129 (1957).

For a discussion on extensions to the Hubbard model which include the J^F terms of (3.10), see

- J.E. Hirsch, Phys. Rev. B **40**, 9061 (1989).

Part II

Wave Functions and Correlations

Illustration by Dick Codor.

4

Ground States of the Hubbard Model

A first step toward understanding a quantum Hamiltonian is to search for its ground state. In the absence of an exact solution, a judicial choice of a family of variational states Ψ^γ, where γ are variational parameters, can be fruitful. The variational theorem states that

$$\frac{\langle \Psi^\gamma | \, \mathcal{H} \, | \Psi^\gamma \rangle}{\langle \Psi^\gamma | \Psi^\gamma \rangle} = E^\gamma \geq E_0, \tag{4.1}$$

where E_0 is the exact ground state energy. A systematic improvement of the ground state energy and wave function can be achieved by minimizing the left-hand side of (4.1) with respect to ever larger families of variational states. The variational approach is conceptually straightforward and avoids the mathematically subtle convergence problems that plague perturbation theories and asymptotic expansions.

On the other hand, we must remember that the variational approach for the ground state order may be grossly misleading. The energy is mostly sensitive to *short-range* correlations. Therefore, within a restricted family of trial states, the state with the lowest energy might have wrong long-range correlations.

In this chapter, our emphasis is on the *magnetic* correlations of the Hubbard model. In Section 4.1 the variational approach is demonstrated using spin density wave Fock states. This provides a variational derivation for Hartree–Fock theory at zero temperature.

The second section presents (without proofs) some relevant theorems for the total spin of the ground states on finite lattices. Subsequent chapters are devoted to the magnetic behavior at half filling, i.e., the Heisenberg model. The reader who is interested in the Hubbard model away from half filling may find Chapter 19 useful. There the two-dimensional doped anti-ferromagnet (relevant to cuprate superconductors) is treated semiclassically using the large spin extension of the $t - J$ model.

4.1   Variational Magnetic States

We consider the Hubbard model in the form

$$\mathcal{H} = \sum_{\mathbf{k},s=\uparrow\downarrow} \epsilon_{\mathbf{k}} c^{\dagger}_{\mathbf{k}s} c_{\mathbf{k}s} + U \sum_i n_{i\uparrow} n_{i\downarrow}. \qquad (4.2)$$

The simplest variational states are Fock states (see Section A.1):

$$|\{n^{\gamma}_{\alpha}\}\rangle, \qquad (4.3)$$

where α label the single-particle states $\{\phi^{\gamma}_{\alpha}\}$, and γ is a set of variational parameters. The expectation values of four Fermi operators in (4.3) factorize as[1]

$$\langle c^{\dagger}_1 c^{\dagger}_3 c_4 c_2 \rangle = \langle c^{\dagger}_1 c_2 \rangle \langle c^{\dagger}_3 c_4 \rangle - \langle c^{\dagger}_1 c_4 \rangle \langle c^{\dagger}_3 c_2 \rangle. \qquad (4.4)$$

Fock states that describe magnetic ordering are called *spin density wave* states. Such states can be constructed by transforming the original electron operators $c_{\mathbf{k}s}$ into magnetic quasiparticles $\alpha_{\mathbf{k}\pm}$ by the canonical transformation

$$\begin{aligned}
\alpha^{\dagger}_{\mathbf{k}+} &= \cos\theta_{\mathbf{k}} c^{\dagger}_{\mathbf{k}\uparrow} + \sin\theta_{\mathbf{k}} c^{\dagger}_{\mathbf{k}+\mathbf{q}\downarrow}, \\
\alpha^{\dagger}_{\mathbf{k}-} &= -\sin\theta_{\mathbf{k}} c^{\dagger}_{\mathbf{k}\uparrow} + \cos\theta_{\mathbf{k}} c^{\dagger}_{\mathbf{k}+\mathbf{q}\downarrow}.
\end{aligned} \qquad (4.5)$$

The variational spin density wave states are given by the family

$$|\Psi^{\mathbf{q},\theta_{\mathbf{k}},\Sigma_F}\rangle = \prod_{\sigma=\pm,\mathbf{k}\in\Sigma^{\pm}_F} \alpha^{\dagger}_{\mathbf{k}\sigma} |0\rangle. \qquad (4.6)$$

The variational parameters of Ψ are the ordering wave vector $\mathbf{q}$, the angles $\theta_{\mathbf{k}}$, and the occupation numbers $n^{\pm}_{\mathbf{k}} = 1, 0$. The latter define the two Fermi surfaces $\Sigma^{\pm}_F$ which enclose $n\mathcal{N}$ occupied states:

$$\begin{aligned}
n &= \frac{1}{\mathcal{N}} \sum_{\sigma,\mathbf{k}} n^{\sigma}_{\mathbf{k}}, \\
n^{\sigma}_{\mathbf{k}} &= \begin{cases} 1 & \mathbf{k} \in \Sigma^{\sigma}_F \\ 0 & \mathbf{k} \notin \Sigma^{\sigma}_F \end{cases}.
\end{aligned} \qquad (4.7)$$

The variational parameters are determined by minimizing the energy of Ψ.

Ψ can support nonzero magnetization in the x–y direction at wave vector $\mathbf{q}$ and also uniform magnetization in the z direction. This can be seen by

---

[1] We do not include here the possibility of anomalous expectation values such as $\langle c^{\dagger} c^{\dagger} \rangle$, which are present in BCS states.

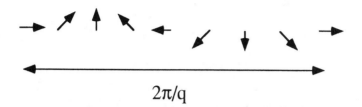

$$\overset{\longleftrightarrow}{2\pi/q}$$

FIGURE 4.1. Spiralling spin order of wave vector q.

evaluating the following spin operators:

$$
\begin{aligned}
\langle S_i^+ \rangle &= \sum_{\mathbf{q}'} e^{-i\mathbf{q}'\mathbf{x}_i} \sum_{\mathbf{k}} \langle c_{\mathbf{k}\uparrow}^\dagger c_{\mathbf{k}+\mathbf{q}'\downarrow} \rangle \\
&= \sum_{\mathbf{q}'} e^{-i\mathbf{q}'\mathbf{x}_i} \sum_{\mathbf{k}} \Bigg(\sin\theta_{\mathbf{k}+\mathbf{q}-\mathbf{q}'} \cos\theta_{\mathbf{k}} \, \langle \alpha_{\mathbf{k}+}^\dagger \alpha_{\mathbf{k}+\mathbf{q}-\mathbf{q}'+} \rangle \\
&\qquad\qquad\qquad\qquad - \sin\theta_{\mathbf{k}} \cos\theta_{\mathbf{k}+\mathbf{q}-\mathbf{q}'} \, \langle \alpha_{\mathbf{k}-}^\dagger \alpha_{\mathbf{k}+\mathbf{q}-\mathbf{q}'-} \rangle \Bigg) \\
&= \mathcal{N} m_{\mathbf{q}} e^{-i\mathbf{q}\mathbf{x}_i} \;, \\
m_{\mathbf{q}} &= \frac{1}{2\mathcal{N}} \sum_{\mathbf{k}} \sin(2\theta_{\mathbf{k}}) \, (n_{\mathbf{k}}^+ - n_{\mathbf{k}}^-).
\end{aligned}
\tag{4.8}
$$

The spin operators in momentum space obey

$$
\langle S^+(\mathbf{q}) \rangle = \langle S^-(-\mathbf{q}) \rangle^* \;.
\tag{4.9}
$$

$m_{\mathbf{q}} \neq 0$ describes *spiralling* spins in the x–y plane with the angles of the spins given by $\mathbf{q} \cdot \mathbf{x}_i$ as shown in Fig. 4.1.

A magnetization in the z direction is described by the order parameter

$$
\begin{aligned}
\langle S_i^z \rangle &= \frac{1}{2} \sum_{\mathbf{q}'} e^{-i\mathbf{q}'\mathbf{x}_i} \sum_{\mathbf{k}} \langle c_{\mathbf{k}\uparrow}^\dagger c_{\mathbf{k}+\mathbf{q}'\uparrow} - c_{\mathbf{k}\downarrow}^\dagger c_{\mathbf{k}+\mathbf{q}'\downarrow} \rangle \\
&= \mathcal{N} m_z \;, \\
m_z &= \frac{1}{2\mathcal{N}} \sum_{\mathbf{k}} \cos(2\theta_{\mathbf{k}}) \, (n_{\mathbf{k}}^+ - n_{\mathbf{k}}^-).
\end{aligned}
\tag{4.10}
$$

The expectation value of the Hubbard interaction, using (4.4), is

$$
\langle U \sum_i n_{i\uparrow} n_{i\downarrow} \rangle = U \sum_i \langle n_{i\uparrow} \rangle \langle n_{i\downarrow} \rangle - \sum_i \langle S_{i\uparrow}^+ \rangle \langle S_{i\downarrow}^- \rangle
$$

$$= \mathcal{N}U\frac{n^2}{4} - \mathcal{N}Um_z^2 - \mathcal{N}Um_{\mathbf{q}}^2. \tag{4.11}$$

The expectation value of the kinetic term is given by

$$T \equiv \langle \sum_{\mathbf{k},s} \epsilon_{\mathbf{k}} c_{\mathbf{k}s}^\dagger c_{\mathbf{k}s} \rangle$$

$$= \sum_{\pm\mathbf{k}} \left[\frac{\epsilon_{\mathbf{k}} + \epsilon_{\mathbf{k+q}}}{2} \pm \left(\frac{\epsilon_{\mathbf{k}} - \epsilon_{\mathbf{k+q}}}{2} \right) \cos(2\theta_{\mathbf{k}}) \right] n_{\mathbf{k}}^\pm. \tag{4.12}$$

By subtracting $-2(\mathcal{N}Um_z^2 - \mathcal{N}Um_{\mathbf{q}}^2)$ from T and adding it to the energy, we obtain the variational expression for the ground state energy:

$$E[\mathbf{q}, \theta_{\mathbf{k}}, \Sigma^\pm] = E^0 + \mathcal{N}m_z^2 + \mathcal{N}m_{\mathbf{q}}^2 + \mathcal{N}n^2/4, \tag{4.13}$$

where

$$E^0 = \sum_{\pm\mathbf{k}} E_{\mathbf{k}}^\pm n_{\mathbf{k}}^\pm. \tag{4.14}$$

$E_{\mathbf{k}}^\pm$ are the "magnetic band energies" defined as[2]

$$E_{\mathbf{k}}^\pm = \frac{\epsilon_{\mathbf{k}} + \epsilon_{\mathbf{k+q}}}{2} \pm \left[\left(\frac{\epsilon_{\mathbf{k}} - \epsilon_{\mathbf{k+q}}}{2} - m_z \right) \cos(2\theta_{\mathbf{k}}) - Um_{\mathbf{q}} \sin(2\theta_{\mathbf{k}}) \right]. \tag{4.15}$$

It is easy to verify that minimization of (4.14) with respect to $m_{\mathbf{q}}$ and m_z leads to (4.8) and (4.10), respectively. Thus, we can treat m_z and $m_{\mathbf{q}}$ as free variational parameters, which are independent of $\cos(2\theta_{\mathbf{k}})$. We must now minimize the functional $E[m_{\mathbf{q}}, m_z, \mathbf{q}, \theta_{\mathbf{k}}, \Sigma_f^\pm]$. It is clear from (4.14) that the minimal $\Sigma_F^\pm$ are those that enclose the *lowest* magnetic band energies.

The variational function $\cos(2\theta_{\mathbf{k}})$ is determined by

$$\frac{\partial E}{\partial \cos(2\theta_{\mathbf{k}})} = \left[\frac{\epsilon_{\mathbf{k}} - \epsilon_{\mathbf{k+q}}}{2} - Um_z + Um_q \cot(2\theta_{\mathbf{k}}) \right] (n_{\mathbf{k}}^+ - n_{\mathbf{k}}^-) = 0, \tag{4.16}$$

which yields the explicit dependence of $\cos(2\theta_{\mathbf{k}})$ on m_z and m_q,

$$\cos(2\theta_{\mathbf{k}}) = \frac{\frac{\epsilon_{\mathbf{k}} - \epsilon_{\mathbf{k+q}}}{2} - Um_z}{\sqrt{\left(\frac{\epsilon_{\mathbf{k}} - \epsilon_{\mathbf{k+q}}}{2} - Um_z \right)^2 + (Um_q)^2}}. \tag{4.17}$$

Inserting (4.17) in (4.15), one obtains the magnetic bands

$$E_{\mathbf{k}}^\pm(m_z, m_{\mathbf{q}}) = \frac{\epsilon_{\mathbf{k}} + \epsilon_{\mathbf{k+q}}}{2} \pm \sqrt{\left(\frac{\epsilon_{\mathbf{k}} - \epsilon_{\mathbf{k+q}}}{2} + Um_z \right)^2 + U^2 m_{\mathbf{q}}^2}. \tag{4.18}$$

---

[2]Here, $E_{\mathbf{k}}$ are variational parameters for the ground state and not true excitations of the model.

The two order parameters, m_z and $m_{\mathbf{q}}$, determine the magnetic bands, the Fermi surfaces, and $E(m_z, m_{\mathbf{q}})$. Minimizing with respect to these parameters yields the coupled equations

$$\frac{dE}{dm_{\mathbf{q}}} = \frac{dE^0}{dm_{\mathbf{q}}} - 2\mathcal{N}Um_q = 0 \ ,$$

$$\frac{dE}{dm_z} = \frac{dE^0}{dm_z} - 2\mathcal{N}Um_z = 0.$$

$$(4.19)$$

Equations (4.19) are known as the zero temperature *Hartree–Fock mean field equations*. They can be solved (numerically, in most cases) for any noninteracting band structure $\epsilon_{\mathbf{k}}$, filling n and interaction U.

From (4.19) we can get an instability criterion for the paramagnetic state with respect to formation of a magnetic state. By assuming $m_{\mathbf{q}} = 0$, and expanding (4.19) to linear order in m_z, we obtain the condition for a uniform magnetization in the z direction ($m_z \neq 0$),

$$\frac{dE^0}{dm_z} = 4U^2 m_z \chi|_{m_z=0} + \mathcal{O}(m_z^2)$$

$$= 2m_z U, \qquad\qquad (4.20)$$

where χ is the *uniform magnetic susceptibility* for noninteracting electrons

$$\chi(0) = \frac{1}{2}\sum_{\mathbf{k}} \frac{dn(\epsilon_{\mathbf{k}})}{d\epsilon_{\mathbf{k}}} = \frac{1}{2}\rho_\uparrow(0), \qquad (4.21)$$

where $n(\epsilon) = (e^{\epsilon/T} + 1)^{-1}$ is the Fermi function, and $\rho_\uparrow(0)$ is the non-interacting, single spin density of states at the Fermi energy. Equation (4.21) yields the famous *Stoner's criterion* for a ferromagnetic instability:

$$2U\chi(0) = 1. \qquad\qquad (4.22)$$

We must also check for an instability toward a spin density wave state of $m_{\mathbf{q}} \neq 0$. By assuming $m_z = 0$, and expanding (4.19) to linear order in $m_{\mathbf{q}}$, we obtain the condition for a solution with $m_{\mathbf{q}} \neq 0$ when

$$\frac{dE^0}{dm_{\mathbf{q}}} = 4U^2 m_{\mathbf{q}}\chi(\mathbf{q})|_{m_{\mathbf{q}}=0} + \mathcal{O}(m_{\mathbf{q}})$$

$$= 2m_{\mathbf{q}}U \ , \qquad\qquad (4.23)$$

where $\chi(\mathbf{q})$ is the *susceptibility* at wave vector $\mathbf{q}$,

$$\chi(\mathbf{q}) = \frac{1}{2}\sum_{\mathbf{k}} \frac{n(\epsilon_{\mathbf{k}+\mathbf{q}}) - n(\epsilon_{\mathbf{k}})}{\epsilon_{\mathbf{k}} - \epsilon_{\mathbf{k}+\mathbf{q}}}. \qquad (4.24)$$

Equation (4.23) also yields a Stoner's criterion for the spin density wave instability at wave vector $\mathbf{q}$:

$$2U\chi(\mathbf{q}) = 1. \tag{4.25}$$

Maximizing $\chi(\mathbf{q})$ therefore determines the ordering wave vector at which the magnetic instability first occurs as we increase the magnitude of U.

By (4.24) we see that while $\chi(0)$ depends on the Fermi surface density of states, $\chi(\mathbf{q})$ at finite $|\mathbf{q}|$ is sensitive to the Fermi surface geometry, in particular its *nesting* properties. "Nesting" refers to the existence of parallel sections on the Fermi surface which are separated by the wave vector $\mathbf{q}_{nest}$. These effects are most dramatic in one dimension, and for the two-dimensional square lattice near half filling. The nesting provides a large number of small energy denominators $|\epsilon_{\mathbf{k}} - \epsilon_{\mathbf{k}+\mathbf{q}_{nest}}| << W$ in the sum (4.24) which enhances $\chi(\mathbf{q}_{nest})$. This divergence may produce a magnetic ground state even for very small U/t!

The magnetic states presented above are commonly used to model metallic ferromagnets (e.g., iron) and spin density wave systems (e.g., chromium). In the Hartree–Fock mean field theory for the excitations, an absence of a gap in the magnetic band structure implies that there are low-lying current carrying excitations. Such systems are called *itinerant magnets*.

All the variational analysis really tells us is that the lowest spin density wave state has lower energy than other trial Fock states, including the nonmagnetic band structure. However, Stoner's criterion is known to *overestimate* the magnetic ordering and *underestimate* quantum disordering effects due to spin fluctuations. The Hartree–Fock equations at finite temperatures find long-range magnetic order even in one and two dimensions. This violates Mermin and Wagner's theorem, which is given in Chapter 6. Therefore, we can only conclude that when the Stoner criterion is satisfied, there is at least short-range magnetic ordering in the Hubbard model.

We have concentrated here only on *magnetic* variational Fock states. In addition, there are other variational Fock states which describe charge density ordering, superconductivity, or a mixture of several different orderings. They are obtained by performing other canonical transformations on the electrons, in analogy to (4.5). For example: we could consider mixing same-spin particles at different momenta to obtain a charge density wave. Mixing particles and holes will yield a superconducting state that breaks gauge symmetry.[3] Each possibility adds variational parameters and mean field equations. Solving for the angles and order parameters is a cumbersome yet straightforward generalization of the calculation we have outlined above.

Before leaving the subject of variational states we should mention an

---

[3] See Schrieffer's book, bibliography of Chapter 3.

important family of non-Fock states, the *Gutzwiller states*

$$\Psi^g \;=\; \prod_i [1 - g n_{i\uparrow} n_{i\downarrow}] \; \Psi^{Fock} \;, \qquad (4.26)$$

where Ψ^{Fock} is any variational Fock state, for instance (4.6). g is an additional parameter which reduces the relative weights of states with doubly occupied sites. $g \to 1$ completely eliminates doubly occupied states. A particularly well-studied member of this family is the Gutzwiller projected Fermi gas in one dimension:

$$\Psi^{GPFG} \;=\; \prod_i (1 - n_{i\uparrow} n_{i\downarrow}) \prod_{|\mathbf{k}| \leq \pi/a} c^\dagger_{\mathbf{k}\uparrow} c^\dagger_{\mathbf{k}\downarrow} |0\rangle \qquad (4.27)$$

Ψ^{GPFG} is a purely spin variational state. It is, in fact, the exact ground state of the Haldane–Shastry Heisenberg model of $S = \frac{1}{2}$ which has antiferromagnetic couplings that decay with distance as $1/r^2$.

The calculation of energy and correlations in Gutzwiller states is far from trivial, since they do not factorize as in (4.4). The perturbative expansion (in g) involves large products of $g n_{i\uparrow} n_{i\downarrow}$. Such calculations have been analytically performed in one dimension by Vollhardt and Metzner.

4.2   Some Ground State Theorems

The total spin and magnetization of the ground state of the Hubbard model obey certain general theorems. Here we shall present some of these theorems without their detailed proofs. The steps of the proofs are instructive in their own right, and the reader is encouraged to learn them directly from the original papers. All the following theorems are restricted to finite lattices of size $\mathcal{N}$ and fixed number of electrons N_e.

Nagaoka's Theorem 4.1 *The ground states of the infinite-U model on a bipartite lattice (see definition 3.1),*

$$\lim_{U \to \infty} \mathcal{H}^{eff} \;=\; -\sum_{ijs} |t_{ij}| P_s c^\dagger_{is} c_{js} P_s, \qquad (4.28)$$

for $N_e = \mathcal{N} - 1$, are the fully polarized ferromagnets with total spin:

$$S \;=\; \frac{1}{2}(\mathcal{N} - 1). \qquad (4.29)$$

One of the (unnormalized) ground states with the maximal magnetization is

$$\Psi^\uparrow \;=\; \sum_i (-1)^i \prod_{i' \neq i} c^\dagger_{i'\uparrow} |0\rangle \;. \qquad (4.30)$$

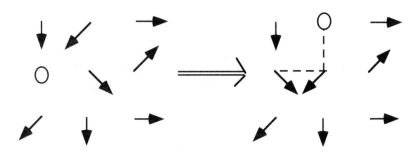

FIGURE 4.2. Two hops of a hole in a spin background.

The ground state energy is $-zt$, where z is the number of nearest neighbors, and their total spin is given by (4.29).

The proof of this theorem was given by Yosuke Nagaoka,[4] and we shall not repeat it here. We just mention that it constructs the dynamical one-particle Green's function for several types of lattices by summing over all closed hopping paths of the hole. As direct corollaries, Theorem 4.1 can be extended to bipartite lattices with one electron above half filling, by applying a particle–hole transformation:

$$c_{is} \rightarrow \eta_i c_{is}^\dagger, \qquad \eta_i = \begin{cases} 1 & i \in A \\ -1 & i \in B \end{cases}, \qquad (4.31)$$

which preserves the Hamiltonian (4.28) and sends $N_e \rightarrow \mathcal{N} + 1$.

Nagaoka ferromagnetism is purely kinetic in origin: it derives from the minimization of the hole's kinetic energy in a purely aligned spin configuration. The argument goes as follows. For an arbitrary spin configuration, a hole that hops around leaves a string of translated spins. Thus, many closed paths do not restore the spins to their initial configuration and are therefore excluded from the kinetic energy. This is depicted in Fig. 4.2. On the other hand, the ferromagnetic spin configuration returns to its original state for any closed loop, and thus its kinetic energy is minimal. More could be found in Nagaoka's original paper and in the bibliography.

This result, however, is not readily extendable to either finite-U or other hole densities. The Nagaoka Hamiltonian (4.28) is somewhat pathological in its excitation spectrum. At $N_e = \mathcal{N}$, the ground state has $2^\mathcal{N}$-fold degeneracy of $\mathcal{N}$ independent spin-half degrees of freedom. This degeneracy is strongly affected by most deviations from the strict conditions of the theorem, be it a finite temperature, a finite-U, or a different filling. Thus, it is

---

[4]See Y. Nagaoka, Phys. Rev. **147**, 392 (1966).

far from clear under which perturbation Nagaoka ferromagnetism survives. The main lesson to learn from this theorem is that mobile holes tend to align the background spins in some range or over some timescale. It suggests that localized ferromagnetic polarons can form around holes in the $t - J$ model (see Chapter 19).

We also mention without proof two theorems by Elliot Lieb.[5]

Lieb's Theorem 4.2 *Consider the positive-U Hubbard model with bipartite hopping parameters t_{ij} (see Definition 3.1),*

$$\mathcal{H} = -\sum_{sij} t_{ij} c_{is}^\dagger c_{js} + \sum_i |U_i| n_{i\uparrow} n_{i\downarrow}. \tag{4.32}$$

$\mathcal{N}_A$ and $\mathcal{N}_B$ are the number of sites on sublattices A and B, respectively. Assuming that the number of electrons is $N_e = \mathcal{N}_A + \mathcal{N}_B$ (half filling), the ground state $|\Psi_0\rangle$ has total spin

$$\mathbf{S}^2|\Psi_0\rangle = S(S+1)|\Psi_0\rangle$$
$$S = \frac{1}{2}|\mathcal{N}_A - \mathcal{N}_B|, \tag{4.33}$$

and also $|\Psi_0\rangle$ is unique up to a trivial $(2S + 1)$-fold rotational degeneracy.

Corollary 4.3 *The ground state of a bipartite lattice with $\mathcal{N}_A = \mathcal{N}_B$ is therefore a total singlet ($S = 0$) and nondegenerate. Thus, there cannot be any level crossing (for the ground state) at finite U.*

In other words, the small-U and large-U ground states at half filling are "adiabatically connected," i.e., under increase of U the ground state evolves continuously from the free electron gas into the ground state of the Heisenberg antiferromagnet as expected from Section 3.2. Chapter 5 describes Marshall's theorem for the Heisenberg model, which closely resembles Lieb's Theorem 4.2. Both theorems use a particular property of the Hamiltonian: its off-diagonal matrix elements are nonpositive in a certain basis. This property results in a ground state that is positive definite on that basis. Consequently, its uniqueness and the value of its total spin can be proven.

We have seen in Section 3.3 that the negative-U Hubbard model at all fillings is related to the positive-U model at half filling. Therefore, Theorem 4.2 is closely related to the following theorem.

Lieb's Theorem 4.4 *Consider the negative-U Hubbard model on a finite lattice ($\mathcal{N} < \infty$),*

$$\mathcal{H} = -\sum_{sij} t_{ij} c_{is}^\dagger c_{js} - \sum_i |U_i| n_{i\uparrow} n_{i\downarrow}, \tag{4.34}$$

---

[5]E. Lieb, Phys. Rev. Lett. **62**, 1201 (1989); Erratum, ibid. **62**, 1927 (1989).

for an even number of electrons with any hopping parameters $t_{ij} = t_{ji}$.
The ground state $|\Psi_0\rangle$ *of (4.34) is a singlet of total spin* $\mathbf{S} = \sum_i \mathbf{S}_i$:

$$\mathbf{S}_{tot}^2 |\Psi_0\rangle = 0, \tag{4.35}$$

and, in addition, it is unique.

This theorem is not surprising for $|U/t| \gg 1$, since the ground state is expected to be a system of locally paired electrons. The uniqueness of the ground state and its spin at smaller $|U/t|$ is a nontrivial result of Theorem 4.4. As discussed in Section 3.3, the negative-U Hamiltonian describes a competition between charge density ordering and superconductivity. Unfortunately, the spin of the ground state cannot help us to distinguish between the possible symmetry breakings in the thermodynamic limit. Those must be established by other means (see Chapter 6).

4.3 Exercises

1. Find the expectation value of $\mathcal{H}$ of (4.2) in the Fermi gas paramagnetic state

$$\Psi^0 = \prod_{s, \mathbf{k} \in \Sigma_F} c_{\mathbf{k}s}^\dagger |0\rangle, \tag{4.36}$$

where Σ_F is the Fermi surface which encloses $\frac{1}{2}\mathcal{N}n$ $\mathbf{k}$ values.

2. Compare the energy of (4.36) to the energy of the fully polarized ferromagnetic state

$$\Psi^\uparrow = \prod_{\mathbf{k} \in \Sigma_F^\uparrow} c_{\mathbf{k}\uparrow}^\dagger |0\rangle, \tag{4.37}$$

where $\Sigma_F^\uparrow$ is the *single spin* Fermi surface which encloses N_e $\mathbf{k}$ states. What is the size of U/t for which the ferromagnetic state is favored? Compare this variational criterion to Stoner's criterion (4.22). Is it stronger or weaker?

3. What is the relation and difference between Stoner's condition (4.22) and Hund's rule for ferromagnetic coupling (see Section 2.1)?

Bibliography

A Green's function derivation of the Hartree–Fock approximation and Stoner's criterion is found in

- S. Doniach and E.H. Sondheimer, *Green's Functions for Solid State Physicists* (Benjamin/Cummings, 1974).

A comprehensive treatise on spin density waves, theory, and experiment is found in

- T. Moriya, *Spin Fluctuations in Itinerant Electron Magnetism* (Springer-Verlag, 1985).

For a recent review on rigorous theorems of the Hubbard model and open problems, see

- E. Lieb, Proceedings of the conference *Advances in Dynamical Systems and Quantum Physics*, Capri (World Scientific, 1993).

Gutzwiller variational states (4.26) were proposed in his classic paper

- M.C. Gutzwiller, Phys. Rev. Lett. **10**, 159 (1963).

Analytical calculations of correlations in Gutzwiller states were performed by

- W. Metzner and D. Vollhardt, Phys. Rev. B **37**, 7382 (1988);
- F. Gebhard and D. Vollhardt, Phys. Rev. B **38**, 6911 (1988).

The Haldane–Shastry model was introduced in

- F.D.M. Haldane, Phys. Rev. Lett. **60**, 635 (1988);
- B.S. Shastry, Phys. Rev. Lett. **60**, 639 (1988),

where the ground state is proven to be the Gutzwiller state Ψ^{GPFG} of (4.27).

5

Ground States of the Heisenberg Model

In Section 3.2, the antiferromagnetic Heisenberg model of spin half emerged as an effective Hamiltonian for Mott insulators. In general, the Heisenberg Hamiltonian is a fundamental model for quantum magnetism, as well as other phenomena that can be effectively described by quantum spin operators.[1] A wide range of concepts and techniques can be learned from studies of its ground state, excitations, and thermodynamic phases.

The Hamiltonian for a lattice of $\mathcal{N}$ spins of size S is

$$
\begin{aligned}
\mathcal{H} &= \mathcal{H}^{zz} + \mathcal{H}^{xy}, \\
\mathcal{H}^{zz} &= \frac{1}{2} \sum_{ij} J_{ij} S_i^z S_j^z, \\
\mathcal{H}^{xy} &= \frac{1}{4} \sum_{ij} J_{ij} \left(S_i^+ S_j^- + S_i^- S_j^+ \right), \\
S_{tot}^z &= \sum_i S_i^z,
\end{aligned}
\tag{5.1}
$$

where $J_{ij} = J_{ji}$, where J_{ij} has the lattice translational symmetry. $\mathcal{H}$ is rotationally invariant since it commutes with all three components of the total spin

$$
\mathbf{S}_{tot} = \sum_i \mathbf{S}_i.
\tag{5.2}
$$

Thus, the eigenstates are labelled by

$$
\begin{aligned}
\Psi &= |S_{tot}, M, \ldots\rangle, \\
M &= -S_{tot}, -S_{tot}+1, \ldots, S_{tot}, \\
S_{tot} &\leq \mathcal{N}S,
\end{aligned}
\tag{5.3}
$$

where M is the eigenvalue of total magnetization S_{tot}^z, and

$$
\mathbf{S}_{tot}^2 = S_{tot}(S_{tot}+1).
\tag{5.4}
$$

---

[1] For example: superconductivity and charge density waves, as shown in Section 3.3.

If the couplings J_{ij} are invariant under lattice translations of i and j, the lattice momentum $\mathbf{k}$ is also a good quantum number.

This chapter concentrates on two aspects of the ground state of the Heisenberg model. In Section 5.1, Marshall's theorems for the total spin and signs of the wave function are proven. Section 5.2 proves gapless behavior in the large $\mathcal{N}$ limit of half-odd integer spin chains.

5.1   The Antiferromagnet

The antiferromagnetic ground state is, in general, far more complicated than the ferromagnetic ground state. The following theorems derive strong conditions on the ground state of *bipartite antiferromagnetic Hamiltonians* (see definition 3.1). The staggered magnetization operator on sublattices A and B is

$$S^{stagg} = \sum_{i \in A} S_i^z - \sum_{i \in B} S_i^z. \tag{5.5}$$

The *Ising configurations* form the basis set

$$\Phi_\alpha = |S, m_1^\alpha\rangle_1 \, |S, m_2^\alpha\rangle_2 \cdots, |S, m_\mathcal{N}^\alpha\rangle_\mathcal{N} \ , \tag{5.6}$$

where $|S, m_i\rangle_i$ denotes an eigenstate of $\mathbf{S}_i^2, S_i^z$ with eigenvalues $S(S+1), m_i$, respectively. The Néel state (which maximizes S^{stagg}) is the Ising configuration

$$\Psi^{N\acute{e}el} = \prod_i |S, \eta_i S\rangle_i \ , \qquad \eta_i = \begin{cases} 1 & i \in A \\ -1 & i \in B \end{cases}. \tag{5.7}$$

Generally, however, $\Psi^{N\acute{e}el}$ *is not an eigenstate of* $\mathcal{H}$. The "spoilers" are the spin flip terms $S_i^+ S_j^-$ in H^{xy}, which connect $\Psi^{N\acute{e}el}$ to other Ising configurations. For our purposes, it is convenient to rotate the spin axes on sublattice B about the z axis, which amounts to the unitary transformation

$$
\begin{aligned}
i &\in B \ , \\
S_i^+ &\rightarrow -\tilde{S}_i^+, \\
S_i^- &\rightarrow -\tilde{S}_i^-, \\
S_i^z &\rightarrow +\tilde{S}_i^z,
\end{aligned}
\tag{5.8}
$$

which transforms the Ising configurations into

$$|S, m_i\rangle_i \rightarrow |S, \tilde{m}_i\rangle_i = \begin{cases} |S, m_i\rangle_i & i \in A \\ (-1)^{(S+m_i)} |S, m_i\rangle_i & i \in B \end{cases}, \tag{5.9}$$

where $\tilde{m}_i$ are eigenvalues of S_i^z and $\tilde{S}_i^z$ on sublattices A and B, respectively. We restrict ourselves to a subspace M and expand all wave functions as

$$\Psi^M = \sum_\alpha f_\alpha^M \tilde{\Phi}_\alpha^M \ , \tag{5.10}$$

where

$$\tilde{\Phi}_\alpha^M = \prod_{\sum_i \tilde{m}_i^\alpha = M} |S, \tilde{m}_i^\alpha\rangle_i. \tag{5.11}$$

Now we present Marshall's theorems, which were extended by Lieb and Mattis (see the bibliography).

Marshall's Theorem 5.1 *Consider the Hamiltonian (5.1), with antiferromagnetic exchanges $J_{ij} \geq 0$ which connect between sublattices A and B (bipartite), such that any two sites on the lattice are connected by a sequence of finite exchanges between intermediary sites. In any allowed M sector, the lowest energy state Ψ_0^M can be chosen to have* positive definite *coefficients in the rotated Ising basis $\tilde{\Phi}_\alpha^M$, i.e.,*

$$\Psi_0^M = \sum_\alpha f_\alpha^M \tilde{\Phi}_\alpha^M , \quad f_\alpha^M > 0, \ \forall\alpha. \tag{5.12}$$

Therefore, by (5.8), the coefficients of Ψ_0^M in terms of the unrotated Ising configurations $|\Phi_\alpha\rangle$ obey the *Marshall sign criterion*

$$\begin{aligned}
\Psi_0^M &= \sum_\alpha (-1)^{\Gamma(\alpha)} f_\alpha^M |\Phi_\alpha\rangle, \\
\Gamma_\alpha &= \sum_{i\epsilon B}(S + m_i^\alpha) \ .
\end{aligned} \tag{5.13}$$

Marshall's Theorem 5.2 *The absolute ground state Ψ_0, for equal size sublattices A and B, is a singlet of total spin*

$$\mathbf{S}_{tot}|\Psi_0\rangle = 0. \tag{5.14}$$

We first must emphasize that not all total singlets obey the Marshall sign criterion (5.12), and conversely not all Marshall sign obeying states are singlets of total spin. The ground state of our Heisenberg antiferromagnet, however, must obey both conditions.

Proof: Using the sublattice rotated operators (5.8), the Hamiltonian (5.1) transforms into

$$\begin{aligned}
\mathcal{H} &= \tilde{\mathcal{H}}^{zz} + \tilde{\mathcal{H}}^{xy}, \\
\tilde{\mathcal{H}}^{zz} &= + \sum_{i\epsilon A, j\epsilon B} |J_{ij}|S_i^z \tilde{S}_j^z, \\
\tilde{\mathcal{H}}^{xy} &= -\frac{1}{2} \sum_{i\epsilon A, j\epsilon B} |J_{ij}|(S_i^+ \tilde{S}_j^- + S_i^- \tilde{S}_j^+).
\end{aligned} \tag{5.15}$$

We note that $\tilde{\mathcal{H}}^{zz}$ is diagonal in the Ising configurations:

$$\tilde{\mathcal{H}}^{zz}|\tilde{\Phi}_\alpha\rangle = e_\alpha|\tilde{\Phi}_\alpha\rangle \ , \tag{5.16}$$

where the superscript M has been dropped off $\tilde{\Phi}^M$, until it is needed. The crucial point is that in this sublattice-rotated representation, $\tilde{\mathcal{H}}^{xy}$ has only *nonpositive* matrix elements:

$$\langle \tilde{\Phi}_\alpha | \tilde{\mathcal{H}}^{xy} | \tilde{\Phi}_\beta \rangle = -|K_{\alpha\beta}|. \tag{5.17}$$

The eigenvalue equation for the coefficients f is

$$-\sum_\beta |K_{\alpha\beta}| f_\beta + e_\alpha f_\alpha = E f_\alpha. \tag{5.18}$$

Consider the trial function

$$\bar{\Psi} = \sum_\alpha |f_\alpha| |\tilde{\Phi}_\alpha\rangle, \tag{5.19}$$

whose energy is

$$\begin{aligned}
\langle \bar{\Psi} | \tilde{\mathcal{H}} | \bar{\Psi} \rangle &= \sum_\alpha e_\alpha |f_\alpha|^2 - \sum_{\alpha\beta} |K_{\alpha\beta}| |f_\alpha| |f_\beta| \\
&\leq \sum_\alpha e_\alpha |f_\alpha|^2 - \sum_{\alpha\beta} |K_{\alpha\beta}| f_\alpha f_\beta = E.
\end{aligned} \tag{5.20}$$

Therefore, it must also be a ground state and satisfy the eigenvalue equation

$$-\sum_\beta |K_{\alpha\beta}| |f_\beta| + e_\alpha |f_\alpha| = E |f_\alpha|. \tag{5.21}$$

Since $\tilde{\Phi}_\alpha$ are not eigenstates[2] of $\tilde{\mathcal{H}}$, e_α are larger than its lowest energy, i.e.,

$$\forall \alpha , \qquad e_\alpha - E > 0. \tag{5.22}$$

Rewriting (5.18 and 5.21) as

$$(e_\alpha - E) f_\alpha = \sum_\beta |K_{\alpha\beta}| f_\beta, \tag{5.23}$$

$$(e_\alpha - E) |f_\alpha| = \sum_\beta |K_{\alpha\beta}| |f_\beta|, \tag{5.24}$$

and taking the absolute value of both sides of the equations, we find

$$\left| \sum_\beta |K_{\alpha\beta}| f_\beta \right| = \sum_\beta |K_{\alpha\beta}| |f_\beta|, \tag{5.25}$$

which implies that

$$f_\beta \geq 0. \tag{5.26}$$

Thus, the trial state and the ground state are one and the same: $\bar{\Psi} = \pm \Psi$. We can prove the stronger condition on f_β as follows.

---

[2]except for the case $M = \mathcal{N}S$, which is a space of one state.

Lemma 5.3

$$\forall \, \beta, \qquad f_\beta > 0. \tag{5.27}$$

Proof: It could be readily verified that any Ising configuration with total magnetization M is connected to all other configurations in the same sector by successive application of the pairwise spin flip operator $K_{\alpha\beta}$. Thus, if f_α vanished for some α, by (5.22) and (5.23), f_β should also vanish for **all** β in the same M sector. Since we need at least one coefficient to be nonzero, all f_β must also be nonzero. This proves Lemma 5.3 and finishes the proof of Theorem 5.1. Q.E.D.

Corollary 5.4 *For any fixed M, Ψ_0 is nondegenerate.*

This follows from (5.27), since there cannot be an orthogonal state to Ψ that has only positive definite coefficients in the M subspace.

In order to prove Theorem 5.2, we first need to show that the ground state of the M sector has the *minimally* possible total spin.

Lemma 5.5

$$(\mathbf{S}_{tot})^2 \, |\Psi^M\rangle \; = M(M+1)|\Psi^M\rangle. \tag{5.28}$$

Proof: We examine the infinite range Hamiltonian on a bipartite lattice with equal number of sites on the two sublattices:

$$\mathcal{H}^\infty = J \sum_{i \epsilon A, j \epsilon B} \mathbf{S}_i \cdot \mathbf{S}_j = J\mathbf{S}_{tot,A} \cdot \mathbf{S}_{tot,B}. \tag{5.29}$$

This model is thus trivially solvable as a two spin problem using

$$\mathbf{S}_{tot} = \mathbf{S}_{tot,A} + \mathbf{S}_{tot,B}. \tag{5.30}$$

The possible values of the total spin operators are

$$S_{tot,A}, S_{tot,B} \; = 0, 1, \ldots, \mathcal{N}S/2, \tag{5.31}$$

which for equal size sublattices means that

$$0 \, \le S_{tot} \, \le \mathcal{N}S, \tag{5.32}$$

and the eigenvalues of (5.29) are

$$E^\infty(S_{tot}) \; = \frac{J}{2} \left[S_{tot}(S_{tot} + 1) - S_{tot,A}(S_{tot,A} + 1) - S_{tot,B}(S_{tot,B} + 1) \right]. \tag{5.33}$$

Since (5.33) monotonically increases with S_{tot}, and $S_{tot} > M$, the ground state in a given M sector has

$$S_{tot} = M \quad . \tag{5.34}$$

Both $\mathcal{H}$ and $\mathcal{H}^\infty$ satisfy the requirements for Marshall's theorem (5.1), and thus their ground states have Marshall signs (5.12). Therefore, their overlap, which involves a sum over positive numbers, cannot vanish, and they must have the same total spin quantum numbers. Thus, we have proven that the lowest energy state of $\mathcal{H}$ in the sector M has spin $S_{tot} = M$. Since all allowed values of total spin $S'_{tot} > M$ have members in the sector with magnetization M, $E(S_{tot})$ obeys

$$\forall \; S'_{tot} \; > \; S_{tot} \;\; \Rightarrow \;\; E(S'_{tot}) \; > \; E(S_{tot}), \tag{5.35}$$

i.e., *the energy is a monotonically increasing function of* S_{tot}. Now consider the sector of $M = 0$. Inequality (5.35) proves that the ground state must have the minimal possible S_{tot}, which by (5.32) is at $S_{tot} = 0$, which proves Theorem 5.2. Q.E.D.

The non-negativity property of the Heisenberg antiferromagnet on biparatite lattices is shared by the Hubbard model at half filling and the negative-U Hubbard model at all fillings as discussed in Section 4.2. It is also a property of the Heisenberg ferromagnet on all lattices.

Corollary 5.6 *For the ferromagnetic model with nonpositive exchanges,*

$$\mathcal{H} = -\frac{1}{2} \sum_{ij} |J_{ij}| \mathbf{S}_i \cdot \mathbf{S}_j, \tag{5.36}$$

the fully ferromagnetic state Ψ^{FM} *is a member of the ground state multiplet, where*

$$\Psi^{FM} = \prod_{i=1}^{\mathcal{N}} |S, S\rangle_i \; . \tag{5.37}$$

The proof, which is given as an exercise, closely follows the proof for the antiferromagnet.

As a member of a degenerate multiplet, Ψ^{FM} *spontaneously breaks the rotational symmetry of the Hamiltonian.* This symmetry breaking is special in that the operator S_{tot}^z commutes with the Hamiltonian and M is a "good quantum number" (i.e., it labels the eigenstates of $\mathcal{H}$). In this respect, the antiferromagnet differs from the ferromagnet. The staggered magnetization does not commute in general with the Hamiltonian. Spontaneous symmetry breaking, however, is still possible for the antiferromagnet in the strict thermodynamic limit ($\mathcal{N} = \infty$), as will be discussed in Section 6.1.

5.2   Half-Odd Integer Spin Chains

Particles with half-odd integer spins, such as electrons, have a peculiar quantum property: their wave function acquires a minus sign under rotations of 2π about any spin axis. When the wave function is localized about any particular direction in spin space, this effect is hardly noticeable. In low dimensions, however, the zero point fluctuations are large. This will be demonstrated by the semiclassical spin wave theory in Chapter 11. Consequently, interference effects due to these negative signs may become important. For example, we shall see that the one-dimensional Heisenberg antiferromagnet has qualitatively different spectra for integer and half odd integer spins. The following theorem by Lieb, Schultz, and Mattis (see bibliography) establishes this difference.

Lieb, Schultz and Mattis Theorem 5.7 *Consider the Heisenberg antiferromagnet $(J > 0)$ on a chain,*

$$\mathcal{H} = J\sum_{j=1}^{\mathcal{N}} \mathbf{S}_j \cdot \mathbf{S}_{j+1} + J\mathbf{S}_{\mathcal{N}} \cdot \mathbf{S}_1, \tag{5.38}$$

where $\mathcal{N}$ is even. For half-odd integer spins, e.g., $S = 1/2, 3/2, \ldots$, there exists an excited state with energy that vanishes as $\mathcal{N} \to \infty$.

Ψ_0 is the ground state of $\mathcal{H}$. Consider the twist operator

$$\mathcal{O}^1 = \exp\left(i\frac{2\pi}{\mathcal{N}}\sum_{j=1}^{\mathcal{N}} jS_j^z\right), \tag{5.39}$$

which rotates each spin incrementally in the x–y plane, such that between the first and last site the spin coordinates are rotated by 2π about the z axis. A "twisted state" is constructed as

$$\Psi_1 \equiv \mathcal{O}^1|\Psi_0\rangle. \tag{5.40}$$

Our proof consists of two propositions:

Proposition 5.8

$$\langle\Psi_1|\Psi_0\rangle = 0 \tag{5.41}$$

and

Proposition 5.9

$$\lim_{\mathcal{N}\to\infty} [\langle\Psi_1|\mathcal{H}|\Psi_1\rangle - E_0] \to 0, \tag{5.42}$$

where E_0 is the energy of Ψ_0.

Proposition 5.8 is proved by showing that the state Ψ_1 is orthogonal to Ψ_0, because their lattice momenta differ by π. Thus, by expanding Ψ_1 in terms of exact eigenstates (which do not include Ψ_0), we could use Proposition 5.9 to prove that there exists at least one excited state with an excitation energy that vanishes in the large lattice limit.

The lattice translation operator U_x translates the spins by one lattice constant:

$$U_x \mathbf{S}_j U_x^{-1} = \mathbf{S}_{j+1},$$
$$U_x \mathbf{S}_\mathcal{N} U_x^{-1} = \mathbf{S}_1 . \qquad (5.43)$$

The Hamiltonian is translationally invariant:

$$[\mathcal{H}, U_x] = 0 , \qquad (5.44)$$

and by Corollary 5.6, Ψ_0 is nondegenerate. Thus, Ψ_0 must be an eigenfunction of U_x, i.e.,

$$U_x \Psi_0 = e^{ik_0} \Psi_0. \qquad (5.45)$$

k_0 is the ground state lattice momentum. The overlap of the twisted state with the ground state is

$$\langle \Psi_0 | \Psi_1 \rangle = \langle \Psi_0 | \mathcal{O}^1 \Psi_0 \rangle$$
$$= \langle \Psi_0 | U_x \mathcal{O}^1 U_x^{-1} | \Psi_0 \rangle. \qquad (5.46)$$

However,

$$U_x \mathcal{O}^1 U_x^{-1} = \mathcal{O}^1 \exp\left(i 2\pi S_1^z\right) \exp\left(-i \frac{2\pi}{\mathcal{N}} S_{tot}^z\right). \qquad (5.47)$$

By Marshall's theorem 5.2, Ψ_0 is a singlet of total spin and thus

$$\exp\left(-i \frac{2\pi}{\mathcal{N}} S_{tot}^z\right) |\Psi_0\rangle = |\Psi_0\rangle. \qquad (5.48)$$

We also use the identity

$$\exp\left(i 2\pi S_1^z\right) = \begin{cases} 1 & S = 0, 1, 2 \ldots \\ -1 & S = \frac{1}{2}, \frac{3}{2}, \frac{5}{2}, \ldots \end{cases} . \qquad (5.49)$$

Thus (5.47) yields

$$U_x \mathcal{O}^1 U_x^{-1} = \begin{cases} \mathcal{O}^1 & S = 0, 1, 2 \ldots \\ -\mathcal{O}^1 & S = \frac{1}{2}, \frac{3}{2} \frac{5}{2}, \ldots \end{cases} . \qquad (5.50)$$

Thus, for half-odd integer S we find that

$$\langle \Psi_0 | \Psi_1 \rangle = -\langle \Psi_0 | \Psi_1 \rangle = 0, \qquad (5.51)$$

which proves Proposition 5.8, Q.E.D.

$\mathcal{O}^1$ is unitary, and thus Ψ_1 is normalized. The energy of the twisted state is

$$\langle \Psi_1 | \mathcal{H} | \Psi_1 \rangle = E_0 + \langle \Psi_0 | \left[\cos(2\pi/\mathcal{N}) - 1\right] \sum_{j=1}^{\mathcal{N}} (S_j^x S_{j+1}^x + S_j^y S_{j+1}^y)$$
$$+ \sin(2\pi/\mathcal{N}) \sum_{j=1}^{\mathcal{N}} (S_j^x S_{j+1}^y - S_j^y S_{j+1}^x) |\Psi_0\rangle. \qquad (5.52)$$

Since Ψ_0 is a singlet of total spin, and therefore rotationally invariant, the expectation value of any operator that is odd under rotations is zero. In particular, a global rotation that takes $S^x \to S^y$ and $S^y \to S^x$ leads to

$$\langle\Psi_0| \sum_{j=1}^{\mathcal{N}} \left(S_j^x S_{j+1}^y - S_j^y S_{j+1}^x\right)|\Psi_0\rangle = 0. \tag{5.53}$$

Expanding the cosine in (5.52), and using the inequality

$$\langle S_j^x S_{j+1}^x + S_j^y S_{j+1}^y\rangle \leq S^2, \tag{5.54}$$

we find that

$$\langle\Psi_1|\mathcal{H}|\Psi_1\rangle - E_0 \leq \frac{2\pi^2 J S^2}{\mathcal{N}} + \mathcal{O}(\mathcal{N}^{-3}), \tag{5.55}$$

which proves Proposition 5.9. Q.E.D.

In the thermodynamic limit, the admixture of the ground state and this low-lying excited state may break a symmetry of $\mathcal{H}$. For example, the ground state and first excited state of the Majumdar–Gosh model[3] are superpositions of dimer configurations that break lattice symmetry. An alternative scenario occurs in the nearest neighbor Heisenberg model of spin half. des Cloizeaux and Pearson have found (from the exact solution) that magnon excitations are *gapless*. These excitations, which are labelled by lattice momentum k, have energies

$$\omega_{\mathbf{k}} = (\pi/2)|\sin k|, \tag{5.56}$$

which vanish in the thermodynamic limit at $k \to 0$ and $k \to \pi$. These excitations are not spin waves [see (11.40)] since they are not small fluctuations about a ground state with broken spin symmetry.

In contrast to half-odd integer spin chains, we shall learn in Chapter 15 that integer spin chains have a gap in the excitation spectrum, called the *Haldane gap*. Historically, des Cloizeaux and Pearson's excitations were known before Haldane predicted the gap for integer spin chains. Nowadays, however, the semiclassical approach makes the integer cases conceptually simpler than the half-odd integer cases. Gaplessness in the latter systems is understood as quantum interference between topological Berry phases, a subject we shall return to in Secion 15.1.

5.3 Exercises

1. Consider the infinite range bipartite Hamiltonian,

$$H = +|J| \sum_{i\in A, j\in B} \mathbf{S}_i \cdot \mathbf{S}_j, \tag{5.57}$$

---

[3] A Heisenberg model with next nearest neighbor interactions of particular strength, see Section 8.2.1.

Find the ground state and ground state energy. Verify Marshall's theorem 5.1.

2. Follow the proof of Marshall's theorem for the antiferromagnet to show that for the ferromagnet (5.36), an expansion in Ising configurations (for fixed magnetization M) yields finite coefficients f_α of the same sign.

3. Find the energy and spin of the ground state of the long-range ferromagnet,

$$\tilde{H} = -|J| \sum_{ij} \mathbf{S}_i \cdot \mathbf{S}_j, \tag{5.58}$$

and use the result with Exercise 1 to prove Corollary 5.6.

Bibliography

Marshall's original paper is

- W. Marshall, Proc. R. Soc. London Ser. **A 232**, 48 (1955).

This chapter follows

- E. Lieb and D.C. Mattis, J. Math. Phys. **3**, 749 (1962),

who extended and strengthened Marshall's theorem.

The proof of gaplessness in half-odd integer spin chains was given by

- E. Lieb, T.D. Schultz, and D.C. Mattis, Ann. Phys. **16**, 407 (1961).

Magnon excitations of the nearest neighbor antiferromagnet were calculated (from Bethe's solution) by

- J. des Cloizeaux and J.J. Pearson, Phys. Rev. **128**, 2131 (1962).

6

Disorder in Low Dimensions

6.1 Spontaneously Broken Symmetry

Spontaneously broken symmetry is a general phenomenon in statistical mechanics and particle physics. It can occur at finite temperatures and when the order parameter does not commute with the Hamiltonian. To focus our discussion, we consider a spin density wave in the z direction defined by the operator

$$S_{\mathbf{q}}^z = \sum_i e^{i\mathbf{q}\cdot\mathbf{x}_i} S_i^z, \tag{6.1}$$

where i labels the lattice sites and $\mathbf{q}$ is the ordering wave vector.

We add a symmetry breaking term to the rotationally invariant Hamiltonian $\mathcal{H}^0$ using an ordering field h:

$$\mathcal{H}(h) = \mathcal{H}_0 - hS_{\mathbf{q}}^z. \tag{6.2}$$

The magnetization per site is

$$
\begin{aligned}
m_{\mathbf{q}}(h) &= \frac{1}{\mathcal{N}Z}\mathrm{Tr}\left[e^{-\mathcal{H}(h)/T} S_{\mathbf{q}}^z\right], \\
Z &= \mathrm{Tr}\left[e^{-\mathcal{H}(h)/T}\right].
\end{aligned}
\tag{6.3}
$$

For convenience, we restrict our discussion to lattices that are symmetric under reflection about the origin, and thus $S_{\mathbf{q}}^z$ in (6.3) is Hermitian and $m_{\mathbf{q}}$ is real.

At finite fields, $m_{\mathbf{q}}(h) > 0$ since the magnetization is induced by the ordering field.

Definition 6.1 *The system has* spontaneously broken symmetry *if it sustains a finite magnetization in the thermodynamic limit even as we take the ordering field to zero from above:*

$$\lim_{h\to 0^+} \lim_{\mathcal{N}\to\infty} m_{\mathbf{q}}(h,\mathcal{N},T) \neq 0. \tag{6.4}$$

The order of limits is crucial in (6.4). Spontaneously broken symmetry can be found in the absence of an ordering field by examining the two point correlation function

$$S^{\alpha\alpha}(\mathbf{q}) = \lim_{h\to 0^+} \frac{1}{ZN} \mathrm{Tr}\left[e^{-\mathcal{H}(h)/T} S^{\alpha}_{\mathbf{q}} S^{\alpha}_{-\mathbf{q}} \right], \quad \alpha \in \{x, y, z\}. \qquad (6.5)$$

In the absence of the ordering field, the correlations are isotropic in spin space, i.e., independent on the direction α. It can be shown (see the exercises) that spontaneously broken symmetry at momentum $\mathbf{q}$ implies *true long-range order* in the two point correlation function:

$$\lim_{N\to\infty} \left[N^{-1} S^{\alpha\alpha}(\mathbf{q}) \right] > 0, \qquad (6.6)$$

or alternatively,

$$\lim_{|\mathbf{x}_i - \mathbf{x}_j|\to\infty} \lim_{N\to\infty} \langle \mathbf{S}_i \cdot \mathbf{S}_j \rangle \neq 0, \qquad (6.7)$$

where $|\mathbf{x}_i - \mathbf{x}_j|$ is the distance between sites i and j. The term "true long-range order" distinguishes (6.7) from *"quasi-long-range order,"* which refers to a power law decay of the correlations at large distances.

6.2   Mermin and Wagner's Theorem

It is expected that for a system with an ordered ground state, thermally excited states reduce the spin correlations at finite temperatures. This happens in classical and quantum systems alike. When the temperature is much higher than the typical coupling energy scale J, we expect the spins to be uncorrelated at large distances and the magnetization $m_{\mathbf{q}}$ to vanish in the absence of an ordering field. This requires a phase transition at some temperature T_c between the ordered and disordered phases.

The ordered ground state of the Heisenberg model breaks a continuous $O(3)$ symmetry in spin space. The spontaneous magnetization could be made to point anywhere on the sphere by choosing the orientation of the ordering field in that direction. The consequences of this continuous symmetry are especially pronounced in lattices of low dimensionality. In fact, it turns out, as we shall see, that for one and two dimensions the phase transition is exactly *at T=0*, i.e., thermal excitations disorder the spins at infinitesimally low temperatures.

In 1966, Hohenberg utilized Bogoliubov's inequality (defined below) to prove the absence of superfluidity at finite temperatures in one and two dimensions (see bibliography). Following essentially the same approach, Mermin and Wagner showed that for short-range spin models in one and two dimensions there cannot be spontaneous ordering at any finite temperature. Their proof, which is presented below, is specialized to the quantum Heisenberg model. It can be readily generalized to a larger class of models with continuous symmetries.

Mermin and Wagner's Theorem 6.2 *For the quantum Heisenberg model*

$$\mathcal{H} = \frac{1}{2} \sum_{ij} J_{ij} \mathbf{S}_i \cdot \mathbf{S}_j - h S_{\mathbf{q}}^z \tag{6.8}$$

with short-range interactions that obey

$$\bar{J} = \frac{1}{2N} \sum_{ij} |J_{ij}| \, |\mathbf{x}_i - \mathbf{x}_j|^2 < \infty, \tag{6.9}$$

there can be no spontaneously broken spin symmetry at finite temperatures in one and two dimensions,

$$\lim_{h \to 0^+} \lim_{N \to \infty} m_{\mathbf{q}}(h, N) = 0. \tag{6.10}$$

A scalar product between any two operators A and B is defined by

$$(A, B) = \frac{1}{Z} \sum_{n,m}' \langle n|A^\dagger|m\rangle \langle m|B|n\rangle \left(\frac{e^{-E_m/T} - e^{-E_n/T}}{E_n - E_m} \right), \tag{6.11}$$

where $\{E_n, |n\rangle\}$ are the energies and eigenstates of $\mathcal{H}$. $\sum'$ excludes termes with $E_n = E_m$. It is easy to see that the weight in the parentheses is nonnegative. Thus (6.11) is a proper scalar product in the product Hilbert space $\{|n\rangle\} \times \{|n\rangle\}$. For $A = B$, (6.11) is the square norm of A, and by (B.27) and (B.10), it is the *susceptibility* of the A operator:

$$(A, A) = \text{Re } R_{AA}(0) = \chi_{AA}. \tag{6.12}$$

Using the inequality $\tanh(x) < x$ it is easy to verify that

$$0 < \left(\frac{e^{-E_m/T} - e^{-E_n/T}}{E_n - E_m} \right) \leq \frac{1}{2T} \left(e^{-E_m/T} + e^{-E_n/T} \right). \tag{6.13}$$

Thus, we obtain the inequality

$$(A, A) \leq \frac{1}{2T} \langle A^\dagger A + A A^\dagger \rangle. \tag{6.14}$$

We use the Cauchy–Schwartz inequality for scalar products and (6.14) to obtain

$$\begin{aligned} |(A, B)|^2 &\leq (A, A)(B, B) \\ &\leq \frac{1}{2T} \langle A^\dagger A + A A^\dagger \rangle (B, B). \end{aligned} \tag{6.15}$$

Let us define an operator C such that

$$B \equiv \left[C^\dagger, \mathcal{H} \right]. \tag{6.16}$$

Thus,

$$
\begin{aligned}
(A, B) &= \langle [C^\dagger, A^\dagger] \rangle, \\
(B, B) &= \langle [C^\dagger, [\mathcal{H}, C]] \rangle,
\end{aligned}
\tag{6.17}
$$

which by (6.15) yields *Bogoliubov's inequality:*

$$
|\langle [C^\dagger, A^\dagger] \rangle|^2 \le \frac{1}{2T} \langle A^\dagger A + AA^\dagger \rangle \langle [C^\dagger, [\mathcal{H}, C]] \rangle.
\tag{6.18}
$$

Now we choose the following operators:

$$
C = S_{\mathbf{k}}^x, \qquad A = S_{-\mathbf{k}-\mathbf{q}}^y,
\tag{6.19}
$$

which yield

$$
\begin{aligned}
\frac{1}{2} \langle A^\dagger A + AA^\dagger \rangle &= \langle S_{\mathbf{k}+\mathbf{q}}^y S_{-\mathbf{k}-\mathbf{q}}^y \rangle \\
&= \mathcal{N} S^{yy}(\mathbf{k}+\mathbf{q}).
\end{aligned}
\tag{6.20}
$$

S^{yy} is the equal-time correlation function as defined in (6.5).
Using (6.19) on the left-hand side of (6.18) we find

$$
\begin{aligned}
\langle [C^\dagger, A^\dagger] \rangle &= \langle [S_{-\mathbf{k}}^x, S_{\mathbf{k}+\mathbf{q}}^y] \rangle \\
&= i \langle S_{\mathbf{q}}^z \rangle = i \mathcal{N} m_{\mathbf{q}}.
\end{aligned}
\tag{6.21}
$$

$F(\mathbf{k})$ is the *double commutator function*

$$
F(\mathbf{k}) = \mathcal{N}^{-1} \langle \left[S_{-\mathbf{k}}^x, [\mathcal{H}, S_{\mathbf{k}}^x] \right] \rangle.
\tag{6.22}
$$

Now the Bogoliubov inequality (6.18) reads

$$
m_{\mathbf{q}}^2 \le \frac{1}{T} S^{yy}(\mathbf{k}+\mathbf{q}) \, F(\mathbf{k}).
\tag{6.23}
$$

$F(\mathbf{k})$ can be bounded by using the Hamiltonian (6.8) as follows:

$$
\begin{aligned}
F(\mathbf{k}) &= hm_{\mathbf{q}} + i\mathcal{N}^{-1} \sum_{ijl} e^{i\mathbf{k}(\mathbf{x}_l - \mathbf{x}_i)} J_{jl} \langle [S_i^x, (S_j^z S_l^y - S_j^y S_l^z)] \rangle \\
&= hm_{\mathbf{q}} + \mathcal{N}^{-1} \sum_{jl} J_{jl} \{ \cos[\mathbf{k}(\mathbf{x}_j - \mathbf{x}_l)] - 1 \} \, \langle S_j^y S_l^y + S_j^z S_l^z \rangle \\
&\le hm_{\mathbf{q}} + \frac{|\mathbf{k}|^2}{2\mathcal{N}} \sum_{jl} |J_{jl}| |\mathbf{x}_j - \mathbf{x}_l|^2 \, |\langle S_j^y S_l^y + S_j^z S_l^z \rangle|.
\end{aligned}
\tag{6.24}
$$

An upper bound

$$
|\langle S_j^y S_l^y + S_j^z S_l^z \rangle| \le |\langle \mathbf{S}_j \cdot \mathbf{S}_l \rangle| \le S(S+1)
\tag{6.25}
$$

is obtained by considering the maximal eigenvalues of $(\mathbf{S}_j + \mathbf{S}_l)^2$ and $\mathbf{S}_i^2$. Using this bound we obtain

$$F(\mathbf{k}) \leq hm_{\mathbf{q}} + S(S+1)\bar{J}\,|\mathbf{k}|^2, \tag{6.26}$$

where $\bar{J}$ was defined in (6.9).

The double commutator function $F(\mathbf{k})$ contains *dynamical information* since it explicitly depends on $\mathcal{H}$. By (6.26), a finite bound for $F(\mathbf{k})$ can be obtained only if $\bar{J}$ is finite, i.e., the interactions decay sufficiently rapidly at large distances.

Combining (6.20), (6.21), (6.23), and (6.26) yields

$$m_{\mathbf{q}}^2 \leq \frac{1}{T}S^{yy}(\mathbf{k}+\mathbf{q})\left[hm_{\mathbf{q}} + S(S+1)\bar{J}|\mathbf{k}|^2\right] \tag{6.27}$$

or

$$S^{yy}(\mathbf{k}+\mathbf{q}) \geq \frac{Tm_{\mathbf{q}}^2}{hm_{\mathbf{q}} + S(S+1)\bar{J}|\mathbf{k}|^2}. \tag{6.28}$$

The sum over $S^{yy}(\mathbf{k})$ is bounded by the size of the spins

$$\mathcal{N}^{-1}\sum_{\mathbf{k}} S^{yy}(\mathbf{k}+\mathbf{q}) = \mathcal{N}^{-1}\sum_{i}\langle(S_i^y)^2\rangle \leq S(S+1). \tag{6.29}$$

By summing over $\mathbf{k}$ on both sides of (6.28) using (6.29) we obtain

$$S(S+1) \geq \frac{Tm_{\mathbf{q}}^2}{(2\pi)^d}\int_0^{\bar{k}}\frac{dk\,k^{d-1}}{hm_{\mathbf{q}} + S(S+1)\bar{J}|\mathbf{k}|^2}, \tag{6.30}$$

where we have transformed the momenta sum into an integral, and estimated a lower bound by choosing $\bar{k}$ to be smaller than the Brillouin zone edge wave vector. This yields an analytical estimate for the right-hand side of (6.30). In one dimension, one obtains

$$S(S+1) \geq \frac{Tm_{\mathbf{q}}^2}{2\pi\sqrt{\bar{J}S(S+1)hm_{\mathbf{q}}}}\arctan\left[\bar{k}\sqrt{\bar{J}S(S+1)/hm_{\mathbf{q}}}\right]$$

$$\xrightarrow{h\to 0} c\frac{Tm_{\mathbf{q}}^{3/2}}{\sqrt{h\bar{J}S(S+1)}} \qquad (d=1), \tag{6.31}$$

where c is a numerical constant. Inverting (6.31), one obtains

$$m_{\mathbf{q}} \leq c\frac{S(S+1)\bar{J}^{1/3}}{T^{2/3}}h^{1/3}. \tag{6.32}$$

In two dimensions, a similar analysis yields

$$S(S+1) \geq \frac{Tm_{\mathbf{q}}^2}{4\pi S(S+1)\bar{J}}\ln\left(1 + \frac{\bar{J}S(S+1)\bar{k}^2}{hm_{\mathbf{q}}}\right), \tag{6.33}$$

which can be inverted to obtain the bound

$$m_{\mathbf{q}} \leq c \, \frac{S(S+1)\bar{J}^{1/2}}{T^{1/2}} \, \frac{1}{\sqrt{|\ln h|}}. \tag{6.34}$$

Equations (6.32) and (6.34) prove that the magnetization must vanish with the ordering field for all finite values of $T, S, \bar{J}$. Since these bounds *do not depend on the lattice size*, this result holds in the thermodynamic limit. Q.E.D.

Since Theorem 6.2 holds for all S, we can deduce that it holds also for the *classical* (S-independent) limit of the Heisenberg model

$$\mathcal{H}^{cl} \equiv \frac{1}{2} \sum_{ij} J_{ij}^{cl} \hat{\Omega}_i \cdot \hat{\Omega}_j \; - h^{cl} \hat{\Omega}_{\mathbf{q}}^z, \tag{6.35}$$

where $\hat{\Omega}$ are c-number unit vectors. The correspondence between (6.8) and (6.35) is given by scaling the parameters:

$$
\begin{aligned}
J_{ij} S(S+1) & \rightarrow J^{cl} \;, \\
hS & \rightarrow h^{cl} \;, \\
m_{\mathbf{q}}/S & \rightarrow m_{\mathbf{q}}^{cl}.
\end{aligned}
\tag{6.36}
$$

Using these parameters eliminates S from inequalities (6.32) and (6.34). Thus, Theorem 6.2 directly applies for the classical model (6.35), Q.E.D.

In $d=3$, the integral in (6.30) does not diverge for $h \rightarrow 0$. Thus, it is possible to satisfy the inequality at some finite temperature. *Therefore, spontaneous symmetry breaking and an ordered phase is expected at low temperatures in three dimensions.*

6.3   Quantum Disorder at $T = 0$

Mermin and Wagner's theorem does not apply at $T = 0$. Let us assume that in the presence of an infinitesimal ordering field, there is a unique ground state $|0\rangle$. At $T = 0$, the expectation value of any operator is

$$\langle A \rangle \; = \langle 0|A|0\rangle. \tag{6.37}$$

The scalar product (6.11) is thus

$$(A, B) \; = \mathrm{Re} \sum_{m \neq 0} \frac{\langle 0|A^\dagger|m\rangle \langle m|B|0\rangle}{E_m - E_0}. \tag{6.38}$$

We choose the operators A, B, and C as in (6.15) and (6.19). The susceptibility is given by

$$\chi^{yy}(\mathbf{k}) \; = \; \mathcal{N}^{-1}(S^y_{-\mathbf{k}}, S^y_{-\mathbf{k}})$$

$$= \mathcal{N}^{-1} \sum_{m \neq 0} \frac{\left|\langle 0|S^y_{\mathbf{k}}|m\rangle\right|^2 + \left|\langle 0|S^y_{-\mathbf{k}}|m\rangle\right|^2}{E_m - E_0}. \tag{6.39}$$

Using (6.39) in the first line of (6.15) and (6.26), we obtain the zero temperature Bogoliubov inequality as

$$\begin{aligned} m^2_{\mathbf{q}} &\leq \chi^{yy}(\mathbf{k}+\mathbf{q})\, F(\mathbf{k}) \\ &\leq \chi^{yy}\left[hm_{\mathbf{q}} + S(S+1)\bar{J}|\mathbf{k}|^2\right]. \end{aligned} \tag{6.40}$$

Inverting the equation and summing over the Brillouin zone, we obtain

$$\frac{m^2_{\mathbf{q}}}{(2\pi)^d} \int_0^{\bar{k}} \frac{dk\, k^{d-1}}{hm_{\mathbf{q}} + S(S+1)\bar{J}|\mathbf{k}|^2} \leq \mathcal{N}^{-1} \sum_{\mathbf{k}} \chi^{yy}(\mathbf{k}). \tag{6.41}$$

Therefore, if

$$\mathcal{N}^{-1} \sum_{\mathbf{k}} \chi^{yy}(\mathbf{k}) = C < \infty, \tag{6.42}$$

and we are in one or two dimensions, $m_{\mathbf{q}}$ must vanish with h in order to satisfy (6.41). In that case there will be no long-range order even at $T = 0$.

We can prove that if there is a gap in the excitation spectrum,

$$E_m - E_0 > \Delta, \quad \forall\ m \neq 0, \tag{6.43}$$

where Δ is independent of $h, \mathcal{N}$, the ground state of the Heisenberg model must be disordered. The gap allows us to bound the susceptibility (6.39) by the correlation function:

$$\chi^{yy}(\mathbf{k}) \leq \frac{2}{\mathcal{N}\Delta} \sum_m \left|\langle 0|S^y_{\mathbf{k}}|m\rangle\right|^2 = 2\frac{S^{yy}(\mathbf{k})}{\Delta}, \tag{6.44}$$

and since $\mathcal{N}^{-1} \sum_{\mathbf{k}} S^{yy} \leq S(S+1)$, condition (6.42) would be satisfied. Interestingly, Δ plays the role of T in inequality (6.14).

As an example, the antiferromagnetic integer spin chains are known to exhibit the "*Haldane gap*" in their excitation spectrum. The Mermin–Wagner theorem at $T = 0$ implies that these models do not possess true long-range order in their ground states. In fact, the nonlinear sigma model analysis predicts that their correlations decay exponentially at large distances (see Section 15.2).

The converse statement (gapless excitations imply long-range order) is false. For example, the spin half Heisenberg antiferromagnet in one dimension has gapless excitations but no long-range order at $T = 0$.

6.4   Exercises

1. Show that since by (6.3)

$$\lim_{h\to 0^+} m_{\mathbf{q}} = -\lim_{h\to 0^-} m_{\mathbf{q}}, \tag{6.45}$$

long-range order requires the divergence of the zero frequency susceptibility

$$\lim_{h,\omega\to 0^+} \chi^{zz}(\mathbf{q},\omega) = \infty. \tag{6.46}$$

Is the opposite true? (i.e., does a diverging susceptibility imply long-range order?)

2. Using the definition of broken symmetry (6.4) and Marshall's theorem, show that there cannot be any broken spin symmetry for the Heisenberg antiferromagnet on a finite bipartite lattice with two equal size sublattices.

3. Show that spontaneously broken symmetry (6.4) implies true long-range order (6.6). *Hint: Use the Cauchy–Schwartz inequality to show that in a finite ordering field* $\mathbf{h_q}$:

$$\langle \mathbf{S_q} \cdot \mathbf{S_{-q}} \rangle \geq \mathcal{N}^2 \left| m_{\mathbf{q}} \right|^2 . \tag{6.47}$$

Show that for a rotationally symmetric Hamiltonian the limit $|\mathbf{h_q}| \to 0$ *of (6.47) exists and leads to (6.6).*

4. Show that (6.47) implies (6.7). *Hint: Use the Cauchy–Schwartz inequality to show that*

$$\langle \mathbf{S}_i \cdot \mathbf{S}_j \rangle \leq S(S+1), \tag{6.48}$$

and prove (6.7) by contradiction.

Bibliography

Bogoliubov's inequality was derived in

- N.N. Bogoliubov, Physik. Abhandl. Sowjetunion **6**, 1, 113, 229 (1962).

The original use of Boboliubov's inequality in establishing absence of long-range order is credited to Hohenberg's paper on superfluidity in low dimensions:

- P.C. Hohenberg, Phys. Rev. 158, 383 (1967).

Mermin and Wagner's theorem was proven in their classic letter

- N.D. Mermin and H. Wagner, Phys. Rev. Lett. **17**, 1133 (1966).

An analogous theorem for relativistic field theories was proven by

- S. Coleman, Comm. Math. Phys. **31**, 259 (1973).

7

Spin Representations

7.1 Holstein–Primakoff Bosons

The spin is a vector operator. In a broken symmetry phase, the expectation value of at least one of its component is nonzero. It is natural to describe the ordered phase in terms of small fluctuations of the spins about their expectation values. This is the content of spin wave theory. In order to carry out that task, Holstein and Primakoff introduced a boson operator b which represents the three spin component operators as

$$
\begin{aligned}
S^+ &= \left(\sqrt{2S - n_b}\right) b, \\
S^- &= b^\dagger \sqrt{2S - n_b}, \\
S^z &= -n_b + S,
\end{aligned}
\tag{7.1}
$$

where $n_b = b^\dagger b$. Using $[b, b^\dagger] = 1$, the operators in (7.1) indeed obey the spin commutation relations

$$
[S^\alpha, S^\beta] = i\epsilon^{\alpha\beta\gamma} S^\gamma,
\tag{7.2}
$$

where the latin indices run over x, y, z, and $\epsilon^{\alpha\beta\gamma}$ is the totally antisymmetric tensor. The Fock space of b is too large. The physical subspace is spanned by the states

$$
\{|n_b\rangle\}_S = \{|0\rangle, |1\rangle \ldots |2S\rangle\}.
\tag{7.3}
$$

Spurious Fock states with $n_b > 2S$ are eliminated by a projector P_S. In this subspace one can verify that $\mathbf{S}^2 = S(S + 1)$. The spin operators (7.1) do not connect the physical to the unphysical subspaces.

The Holstein–Primakoff representation is useful to describe the broken symmetry phases of the quantum Heisenberg model. The square root factors in (7.1) can be expanded in powers of $1/S$,

$$
\sqrt{2S - n_b} = \sqrt{2S}\left(1 - \frac{n_b}{4S} - \frac{n_b^2}{32S^2} \cdots\right),
\tag{7.4}
$$

which is a semiclassical expansion of the spin fluctuations about the z direction. By inserting (7.4) into the Hamiltonian and keeping up to quadratic terms, one obtains a noninteracting *"spin wave"* Hamiltonian. The higher-order terms introduce interactions between spin waves. The truncation will couple the physical and unphysical subspaces. Nevertheless, low-order spin wave approximation has been found to be remarkably successful in certain ordered phases of the Heisenberg ferromagnet and antiferromagnet. In the bibliography, we refer the reader to some milestone papers on this subject. In Chapter 11, spin wave theory is derived in two ways: from the spin coherent states path integral and by expanding Holstein and Primakoff's operators.

7.2   Schwinger Bosons

The *symmetric* phases of the Heisenberg model are easier to describe using representations in which the rotational invariance of the Hamiltonian is manifested. Two *Schwinger bosons*, a and b, represent the spin operators as follows:

$$
\begin{aligned}
S^x + iS^y &= a^\dagger b, \\
S^x - iS^y &= b^\dagger a, \\
S^z &= \frac{1}{2}(a^\dagger a - b^\dagger b) .
\end{aligned}
\tag{7.5}
$$

It is easy to verify that the spin components, as defined above, satisfy (7.2). The spin magnitude S defines the physical subspace

$$
\{|n_a, n_b\rangle \ : \ n_a + n_b = 2S\}.
\tag{7.6}
$$

This subspace is given by the projector (or "constraint") P_S,

$$
P_S \left(a^\dagger a + b^\dagger b - 2S\right) = 0.
\tag{7.7}
$$

In the projected subspace, the spin magnitude is well defined, i.e.,

$$
\mathbf{S}^2 \, P_S = S(S+1) \, P_S.
\tag{7.8}
$$

In Fig. 7.1 the subspace projected by P_S is depicted in the Fock space n_a, n_b.

The spin states are given by

$$
|S, m\rangle = \frac{(a^\dagger)^{S+m}}{\sqrt{(S+m)!}} \frac{(b^\dagger)^{S-m}}{\sqrt{(S-m)!}} |0\rangle,
\tag{7.9}
$$

where $|S, m\rangle$ label the $\mathbf{S}^2$ and S^z eigenvalues, and $|0\rangle$ is the Schwinger bosons vaccuum. For example: the spin-$\frac{1}{2}$ states are given in the second quantized notation as

$$
\begin{aligned}
|\uparrow\rangle &= a^\dagger |0\rangle , \\
|\downarrow\rangle &= b^\dagger |0\rangle.
\end{aligned}
\tag{7.10}
$$

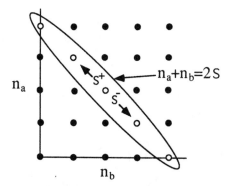

FIGURE 7.1. Projected subspace of spin S in the Schwinger bosons Fock space.

Schwinger bosons are useful for calculating matrix elements of spin operators. Since they do not contain square roots (in contrast to Holstein–Primakoff bosons), the matrix elements of spin operators between Fock states factorize as for free bosons. However, this does not necessarily simplify the calculation of spin correlations in non-Fock wave functions. The local constraints on the Hilbert space (7.6) introduce correlations between a and b occupation numbers.

A generalization of representation (7.5) to N flavors defines the generators of SU(N) as generalized "spins"; see Chapter 16. The large N generalizations are amenable to simple mean field theories which are controlled by the small parameter $1/N$. These will be reviewed in Chapters 17 and 18. The large-N mean field theories are described by effectively noninteracting Bose quasiparticles where correlations between different flavors are ignored.

The Schwinger bosons (SB) and Holstein–Primakoff (HP) bosons (7.1) are closely related. By eliminating the a boson using the constraint (7.7), the correspondence is found to be

$$
\begin{array}{ccc}
\text{SB} & & \text{HP} \\
b & \leftrightarrow & b, \\
a & \leftrightarrow & \sqrt{2S - n_b}.
\end{array}
\tag{7.11}
$$

While Schwinger bosons provide a symmetric representation in spin space, Holstein–Primakoff bosons single out the S^z direction, which defines their vacuum. Therefore, the two representations lend themselves to different approximation schemes: HP for the broken symmetry phases and SB for the symmetric phases.

7.2.1 SPIN ROTATIONS

Since spin operators are generators of the SU(2) group of transformations. The group members $\mathcal{R}$ are parametrized by three Euler angles ϕ, θ, χ

$$\mathcal{R} = e^{i\phi S^z} e^{i\theta S^y} e^{i\chi S^z}. \tag{7.12}$$

$\mathbf{S}$ are normal bilinear operators. The Schwinger boson creation operators transform as *vectors* in SU(2)[1]:

$$
\begin{aligned}
\begin{pmatrix} a^\dagger \\ b^\dagger \end{pmatrix}' &= \mathcal{R} \begin{pmatrix} a^\dagger \\ b^\dagger \end{pmatrix} \mathcal{R}^{-1} \\
&= e^{i\frac{1}{2}\chi\sigma^z} e^{i\frac{1}{2}\theta\sigma^y} e^{i\frac{1}{2}\phi\sigma^z} \begin{pmatrix} a^\dagger \\ b^\dagger \end{pmatrix} \\
&= \begin{pmatrix} u\, e^{i\chi/2} & v\, e^{i\chi/2} \\ -v^*\, e^{-i\chi/2} & u^*\, e^{-i\chi/2} \end{pmatrix} \begin{pmatrix} a^\dagger \\ b^\dagger \end{pmatrix},
\end{aligned}
\tag{7.13}
$$

where the functions u and v are defined as

$$
\begin{aligned}
u(\theta, \phi) &= \cos(\theta/2)e^{i\phi/2}, \\
v(\theta, \phi) &= \sin(\theta/2)e^{-i\phi/2}.
\end{aligned}
\tag{7.14}
$$

7.3 Spin Coherent States

Spin coherent states are a family of spin states created by applying the rotation operator $\mathcal{R}$ of (7.12) to the maximally polarized state $|S, S\rangle$:

$$
\begin{aligned}
|\hat{\Omega}\rangle &= \mathcal{R}(\chi, \theta, \phi)|S, S\rangle \\
&= e^{iS^z\phi} e^{iS^y\theta} e^{iS^z\chi}|S, S\rangle,
\end{aligned}
\tag{7.15}
$$

where the unit vector

$$\hat{\Omega} = (\sin\theta\cos\phi, \sin\theta\sin\phi, \cos\theta) \tag{7.16}$$

parametrizes the spin coherent state. We have the freedom to define χ arbitrarily. This is a *gauge freedom*, which we can eliminate by fixing χ. The two independent angles are θ, the "latitude," and ϕ, the "longitude," which are defined within the domains

$$\{ \theta \in [0, \pi], \ \phi \in [-\pi, \pi) \}. \tag{7.17}$$

---

[1]See (A.17) in Appendix A.

Using (7.13), Schwinger bosons are transformed into the rotated frame $a \to a'_{\theta,\phi}$, which yields an explicit representation of the coherent states as

$$
\begin{aligned}
|\hat{\Omega}\rangle &= e^{iSx} \frac{(a^{\dagger\prime})^{2S}}{\sqrt{(2S)!}} |0\rangle = e^{iSx} \frac{(ua^{\dagger} + vb^{\dagger})^{2S}}{\sqrt{(2S)!}} |0\rangle \\
&= e^{iSx} \sqrt{(2S)!} \sum_m \frac{u^{S+m} v^{S-m}}{\sqrt{(S+m)!(S-m)!}} |S,m\rangle, \quad (7.18)
\end{aligned}
$$

where u and v were previously defined in (7.14).

Using (7.18), we can evaluate the overlap of any two coherent states:

$$
\begin{aligned}
\langle \hat{\Omega} | \hat{\Omega}' \rangle &= e^{-iS(x-x')} (u^* u' + v^* v')^{2S} \\
&= \left(\frac{1 + \hat{\Omega} \cdot \hat{\Omega}'}{2} \right)^S e^{-iS\psi} \\
\psi &= 2 \arctan \left[\tan \left(\frac{\phi - \phi'}{2} \right) \frac{\cos[\frac{1}{2}(\theta + \theta')]}{\cos[\frac{1}{2}(\theta - \theta')]} \right] + x - x', \quad (7.19)
\end{aligned}
$$

where χ, χ' depend on the gauge convention.

The coherent states are not orthogonal, as seen in (7.19). Now we show that they span the space of states of spin S. The measure of integration over the group parameters is defined as

$$
\frac{2S+1}{4\pi} d\hat{\Omega} = \frac{2S+1}{4\pi} d\theta \sin \theta \, d\phi. \quad (7.20)
$$

This is the *Haar measure* of the SU(2) Lie group. Using this measure, the *resolution of identity* is provided by the integral[2]

$$
\begin{aligned}
\frac{2S+1}{4\pi} &\int d\hat{\Omega} \, |\hat{\Omega}\rangle\langle\hat{\Omega}| \\
&= \frac{(2S+1)}{2} \int_{-1}^{1} d\cos\theta \sum_m \left(\frac{1 + \cos\theta}{2} \right)^{S+m} \left(\frac{1 - \cos\theta}{2} \right)^{S-m} \\
&\quad \times \frac{(2S)!}{(S+m)!(S-m)!} |S,m\rangle\langle S,m| \\
&= \sum_m |S,m\rangle\langle S,m| = I. \quad (7.21)
\end{aligned}
$$

Thus the coherent states form *an overcomplete basis*.

Another useful relation can be proven in the same fashion,[3]

$$
\frac{(S+1)(2S+1)}{4\pi} \int d\hat{\Omega} \, \hat{\Omega}^\alpha \, |\hat{\Omega}\rangle\langle\hat{\Omega}| = S^\alpha, \quad \alpha = x, y, z. \quad (7.22)
$$

---

[2]See Subsection 7.3.1 for the evaluation of the θ integrals.
[3]See Exercise 3.

The relations above are easily extended to many spins. Consider a lattice of spins on $\mathcal{N}$ sites, which are labelled by i. The many spin coherent states are products of the single spin states:

$$|\hat{\Omega}\rangle = \prod_{i=1}^{\mathcal{N}} |\hat{\Omega}_i\rangle. \tag{7.23}$$

Their overlap is

$$\langle \hat{\Omega} | \hat{\Omega}' \rangle = \prod_i \left(\frac{1 + \hat{\Omega}_i \cdot \hat{\Omega}_i'}{2} \right)^S e^{-iS \sum_i \psi[\hat{\Omega}_i, \hat{\Omega}_i']}. \tag{7.24}$$

The resolution of the identity in the product space is

$$\int \prod_i \left(\frac{2S+1}{4\pi} d\hat{\Omega}_i \right) |\hat{\Omega}\rangle\langle\hat{\Omega}| = I. \tag{7.25}$$

The spin correlations of any wave function Ψ can be computed by the integral:

$$\frac{\langle \Psi | \mathbf{S}_i \cdot \mathbf{S}_j | \Psi \rangle}{\langle \Psi | \Psi \rangle} = \frac{(S+1-\delta_{ij})(S+1)}{Z} \int \prod_i d\hat{\Omega}_i \left| \Psi[\hat{\Omega}] \right|^2 \hat{\Omega}_i \cdot \hat{\Omega}_j,$$

$$Z = \int \prod_i d\hat{\Omega}_i \left| \Psi[\hat{\Omega}] \right|^2, \tag{7.26}$$

where we have used (7.22). The representation $\Psi[\hat{\Omega}] = \langle \Psi | \hat{\Omega} \rangle$ is continuous (like Schrödinger wave functions), and in Exercise 4 it is shown that all spin operators are represented by differential operators in the variables $u(\theta, \phi)$ and $v(\theta, \phi)$. Spin coherent states can be used to evaluate the trace of any operator. If $|n\rangle$ is any orthonormal basis, then

$$\begin{aligned} \text{Tr}\,\mathcal{O} &= \sum_n \langle n | \mathcal{O} | n \rangle \\ &= \left(\frac{2S+1}{4\pi} \right)^2 \int d\hat{\Omega} \int d\hat{\Omega}' \sum_n \langle n | \hat{\Omega} \rangle \langle \hat{\Omega} | \mathcal{O} | \hat{\Omega}' \rangle \langle \hat{\Omega}' | n \rangle \\ &= \frac{2S+1}{4\pi} \int d\hat{\Omega} \langle \hat{\Omega} | \mathcal{O} | \hat{\Omega} \rangle. \end{aligned} \tag{7.27}$$

The coherent states elucidate the correspondence between classical and quantum spins. The classical limit is achieved by sending $S \to \infty$. In that limit, according to (7.24), the overlap of different coherent states vanishes exponentially with S. The expectation values of spin operators are functions of unit vectors, exactly like classical spins. *Quantum effects are therefore associated with the nonorthogonality of coherent states, which also implies a finite width of* $\Psi[\hat{\Omega}]$ *in* $\hat{\Omega}$ *space.*

7.3.1   THE θ INTEGRALS

We define the θ integral in (7.21) as

$$I_{S,m} = \frac{1}{2}\int_0^\pi d\theta \sin\theta \left(\frac{1+\cos\theta}{2}\right)^{S+m}\left(\frac{1-\cos\theta}{2}\right)^{S-m}. \qquad (7.28)$$

We change integration variables by defining

$$x = \frac{\cos\theta + 1}{2}, \qquad (7.29)$$

and show that

$$\begin{aligned}
I_{S,m} &= \int_0^1 dx\, x^{S+m}(1-x)^{S-m} \\
&= (S+m)!(S-m)!/(2S+1)! \ . \qquad (7.30)
\end{aligned}$$

This is verified by constructing the generating function

$$f_S(z) = \sum_{n=0}^{2S} \frac{(2S)!}{(2S-n)!n!} I_{S,n-S}\, z^n. \qquad (7.31)$$

Then, we insert (7.30) into (7.31) and use the binomial expansion to obtain

$$\begin{aligned}
(2S+1)f_S(z) &= (z^{2S+1}-1)/(z-1) \\
&= 1 + z + \ldots + z^{2S} \ . \qquad (7.32)
\end{aligned}$$

But by (7.31), $I_{S,m}$ is given by the coefficient of z^{m+S} in $f_S(z)$, which proves (7.30).

7.4   Exercises

1. Verify, directly from the definition (7.15), that $|\hat\Omega\rangle$ is an eigenstate of the spin component in the $\hat\Omega$ direction:

$$\hat\Omega\cdot\mathbf{S}|\hat\Omega\rangle = S|\hat\Omega\rangle. \qquad (7.33)$$

Hint: Use the $O(3)$ (vector) transformation properties of the three spin operators (S^x, S^y, S^z).

2. Using (7.33), show that the expectation value of the Heisenberg model in a coherent state is

$$H[\hat\Omega] = \langle\hat\Omega|\mathcal{H}|\hat\Omega\rangle = \frac{S^2}{2}\sum_{i,j} J_{ij}\hat\Omega_i\cdot\hat\Omega_j. \qquad (7.34)$$

Hint: For each of the spins, write its components in the $\hat\Omega$ direction and two transverse directions.

3. Following the derivation of (7.21), prove identity (7.22).

4. Show that for an arbitrary wave function of a single spin, the following relations hold:

$$\begin{aligned}
\langle\Psi|S^+|\hat{\Omega}\rangle &= v\partial_u\Psi(\hat{\Omega}) \ , \\
\langle\Psi|S^-|\hat{\Omega}\rangle &= u\partial_v\langle\Psi|\hat{\Omega}\rangle \ , \\
\langle\Psi|S^z|\hat{\Omega}\rangle &= \frac{1}{2}(u\partial_u - v\partial_v)\Psi(\hat{\Omega}).
\end{aligned} \tag{7.35}$$

5. You are given a differential operator

$$\mathcal{T} = \sum_{klj} T_{klj}\partial_u^k\partial_v^l u^{k+j}v^{l-j}, \tag{7.36}$$

whose matrix elements between any two spin-S wave functions are

$$\langle\Psi|\mathcal{T}|\Psi'\rangle = \frac{2S+1}{4\pi}\int d\hat{\Omega} \ \Psi^*(\hat{\Omega}) \ \left[\mathcal{T}_{u,v}\Psi'(\hat{\Omega})\right]. \tag{7.37}$$

Prove that one can evaluate (7.37) by substituting

$$\langle\Psi|\mathcal{T}|\Psi'\rangle = \frac{2S+1}{4\pi}\int d\hat{\Omega} \ \Psi^*(\hat{\Omega}) \ T[u^*,v^*,u,v], \Psi'(\hat{\Omega}), \tag{7.38}$$

where

$$T[u^*,v^*,u,v] = \sum_{klj} C_{kl}T_{klj}u^{*k}v^{*l}u^{k+j}v^{l-j} \tag{7.39}$$

and

$$C_{kl} = \prod_{m=2}^{k+l+1} (2S+m). \tag{7.40}$$

Hint: expand Ψ, Ψ' into binomials of u,v, integrate first over $d\phi$, and then over $d\theta$ for both (7.37) and (7.40).

Bibliography

Holstein–Primakoff operators were defined in

- T. Holstein and H. Primakoff, Phys. Rev. **58**, 1908 (1940).

Schwinger bosons are defined and used for evaluating Clebsch–Gordan coefficients in

- J. Schwinger, *Quantum Theory of Angular Momentum*, edited by L. Biedenharn and H. Van Dam (Academic, New York, 1965).

Spin coherent states and their generalizations are found in

- A. Perelomov, *Generalized Coherent States and Their Applications* (Springer-Verlag, New York, 1986);

- J.R. Klauder and B. Skagerstam, *Coherent States* (World Scientific, 1985).

Spin coherent states and their relation to the semiclassical approximation is described in

- J. Klauder, Phys. Rev. D **19**, 2349 (1979).

Their use for the quantum Heisenberg ferromagnet is found in

- F.D.M. Haldane, Phys. Rev. Lett. **57**, 1488 (1986).

8

Variational Wave Functions and Parent Hamiltonians

We do not know the ground state for most nonferromagnetic Heisenberg models. Even those that are known in analytical form, such as the Bethe solution in one dimension, require numerical computations for their spin correlations. Variational wave functions provide educated guesses for the ground state. Physical insight can be gained by a variational calculation, where the energy is minimized with respect to a chosen set of parameters. The variational approach takes us beyond the regimes of any particular expansion scheme—semiclassical, large N, or others. As shown for the Hubbard model (see Section 4.1), the primary advantage of the variational approach is that it is conceptuallly simple and clear.

Here, we shall tie the variational wave functions to the concept of a "parent Hamiltonian." This is the Hamiltonian for which a particular wave function happens to be the exact ground state. Although the Hamiltonian may differ from the physical model by extra interactions, it serves to enhance our understanding of the latter. It brings to light the relation between interactions and ground state correlations. Extension of the parent Hamiltonian method to understanding the excitations will be the topic of Chapter 9.

As shown in the bibliography, much of the philosophy of the variational states and parent Hamiltonians is shared with other highly correlated quantum systems. In particular, the methods explained in this chapter are closely analogous to well-known treatments of the fractional quantum Hall effect.

8.1 Valence Bond States

The *valence bond states* are variational wave functions for antiferromagnetic Heisenberg models. They have been studied extensively in the context of quantum magnetism and high-temperature superconductivity. Their general form is

$$|\{c_\alpha\}, S\rangle = \sum_\alpha c_\alpha |\alpha\rangle , \qquad (8.1)$$

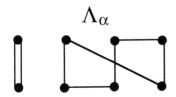

FIGURE 8.1. A configuration of valence bonds for $S = 1$.

where c_α are variational parameters and

$$|\alpha\rangle = \prod_{(ij)\in\Lambda_\alpha} \left(a_i^\dagger b_j^\dagger - b_i^\dagger a_j^\dagger\right)|0\rangle . \tag{8.2}$$

a_i, b_i are Schwinger bosons[1] on site i. Λ_α is a particular configuration of bonds (ij) on the lattice, as, e.g., depicted in Fig. 8.1. The condition on Λ_α is that precisely $2S$ bonds will emanate from each site. In certain cases, the sum in (8.1) is dominated by a finite number of configurations in the large lattice limit. The cases where there are macroscopically many configurations in (8.1) have been denoted *"resonating valence bonds states"* (RVB) by Anderson.

By (7.13), it is easy to verify that all *bond operators* $a_i^\dagger b_j^\dagger - b_i^\dagger a_j^\dagger$ are invariant under global spin rotations. Therefore, $|\{c_\alpha\}, S\rangle$ is a singlet of total spin. A special class of valence bond states is given by

$$|\hat{u}, S\rangle = \sum_\alpha \left(\prod_{(ij)\in\alpha} u_{ij}\right)|\alpha\rangle . \tag{8.3}$$

There are up to $\mathcal{N}(\mathcal{N}-1)/2$ independent variational parameters u_{ij}, where $\mathcal{N}$ is the number of lattice sites.

If the bond parameters are bipartite and positive

$$u_{ij} = \begin{cases} u_{ij} > 0 & i \in A \text{ and } j \in B \\ 0 & i, j \text{ on the same sublattice ,} \end{cases} \tag{8.4}$$

then (8.3) obeys the Marshall sign criterion (5.13) as required for the ground state of a bipartite Heisenberg antiferromagnet.

---

[1]See Section 7.2.

A *Schwinger boson mean field state* is defined as

$$|\hat{u}\rangle = \exp\left[\frac{1}{2}\sum_{ij} u_{ij}\left(a_i^\dagger b_j^\dagger - b_i^\dagger a_j^\dagger\right)\right]|0\rangle, \qquad (8.5)$$

where $u_{ij} = -u_{ji}$. Such states are generated by the Schwinger boson mean field theory of the Heisenberg antiferromagnet, which will be reviewed in Chapter 18. $|\hat{u}\rangle$ includes contributions of different spin sizes at all sites, and is therefore not a bona fide spin state. It can be transformed by a Bogoliubov transformation on the Schwinger bosons, into a factorizable Fock state (which is in fact a vaccuum of the transformed bosons). As seen in the exercises, the correlations of the mean field states can be evaluated analytically. The valence bond state $|\hat{u}, S\rangle$ can be constructed from (8.5) by projecting it with P_S of (7.7):

$$|\hat{u}, S\rangle = P_S|\hat{u}\rangle. \qquad (8.6)$$

The projected state, however, is not factorizable, due to the correlations introduced by P_S.

It is much easier to evaluate the correlations in $|\hat{u}\rangle$ than in $|\hat{u}, S\rangle$. The methods that can be used to calculate the spin correlations of valence bond states are diverse. There are numerical methods, which involve either combinatorial computations or a Monte Carlo sampling of the wave function. Then there are exact calculations using the spin coherent states representation (see Subsection 8.3.1). Also, there is a $1/N$ expansion, where N is the number of Schwinger boson flavors. The overall magnitude of the matrix $\{\hat{u}\}$ is fixed to yield an average of $2S$ bosons per site. The spin correlations of the mean field state $|\hat{u}\rangle$ are the zeroth-order approximation in this expansion. This approximation can be systematically improved by a $1/N$ expansion of the constrained generating functional, in close analogy to the large-N expansion of the partition function in Chapter 16. Samples of these approaches are listed in the bibliography.

Certain valence bond states are exact ground states of known "parent" Hamiltonians. In the following, we describe the *projectors* technique, which is generally useful for constructing parent Hamiltonians.[2]

8.2 $S = \frac{1}{2}$ States

$S = \frac{1}{2}$ states at any site will be denoted by the *spinor states*

$$|\uparrow_i\rangle \equiv a_i^\dagger|0\rangle, \quad |\downarrow_i\rangle \equiv b_i^\dagger|0\rangle . \qquad (8.7)$$

---

[2]The projectors method was used by Haldane to justify Laughlin's wave function for the fractional quantum Hall effect, see bibliography.

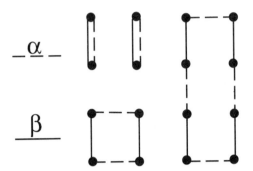

FIGURE 8.2. Overlap of two valence bond configurations $(S = \frac{1}{2})$.

Any bipartite valence bond configuration (8.2) that obeys the Marshall sign can be written in a normalized form as a product of *singlet bonds*:

$$|\alpha\rangle_{S=\frac{1}{2}} = \prod_{\substack{(ij)\in\Lambda_\alpha}}^{i\in A, j\in B} (|\uparrow_i\rangle|\downarrow_j\rangle - |\downarrow_i\rangle|\uparrow_j\rangle)/\sqrt{2}. \qquad (8.8)$$

The spin correlations of (8.8) are simply

$$\langle \mathbf{S}_i \mathbf{S}_j \rangle = \begin{cases} \frac{3}{4} & i = j \\ -\frac{3}{4} & (ij) \in \Lambda_\alpha \\ 0 & (ij) \notin \Lambda_\alpha \end{cases}. \qquad (8.9)$$

Thus if the bonds in Λ_α are of short range, $|\alpha\rangle_{S=\frac{1}{2}}$ is a disordered *"spin liquid"* state. Different valence bond configurations are not orthogonal since their finite overlap is given by[3]

$$\begin{aligned} \langle \alpha | \beta \rangle &= \prod_{l \in \langle\alpha|\beta\rangle} 2^{1-L_l/2} \\ &= 2^{N_L^{\langle\alpha\beta\rangle} - \mathcal{N}}, \end{aligned} \qquad (8.10)$$

where the first line is a product over all loops l of length L_l found in the overlap of the two configurations. The overlap can be depicted by overlaying the two valence bond configurations as shown in Fig. 8.2. Two identical bonds in α and β produce a loop of length $L = 2$. In the second line of (8.10) N_L is the total number of loops and $\mathcal{N}$ is the number of sites on the lattice.

---

[3]See Sutherland.

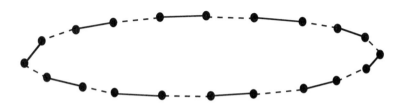

FIGURE 8.3. The two dimer states $|d\rangle_\pm$ depicted with solid and dashed lines, respectively.

Since $|\alpha\rangle$ contains one bond per site, and the coordination number of all simple lattices is equal to or larger than two, $|\alpha\rangle$ breaks lattice translational symmetry. This symmetry can be restored in (8.1) by summing over α.

Henceforth we restrict ourselves to valence bond states with nearest neighbor bonds or *"dimers."* We discuss the one and two-dimensional cases separately.

8.2.1 THE MAJUMDAR–GHOSH HAMILTONIAN

Majumdar and Ghosh introduced the Hamiltonian

$$H^{MG} = \frac{4|K|}{3} \sum_{i=1}^{\mathcal{N}} \left(\mathbf{S}_i \cdot \mathbf{S}_{i+1} + \frac{1}{2} \mathbf{S}_i \cdot \mathbf{S}_{i+2} \right) + \frac{1}{2}\mathcal{N} , \qquad (8.11)$$

where i labels the sites of a one-dimensional chain with even number of sites and $\mathbf{S}_{\mathcal{N}+1} = \mathbf{S}_1$. Below we shall see that H^{MG} is the *parent Hamiltonian* of the two dimer states, which are depicted in Fig. 8.3,

$$|d\rangle_\pm = \prod_{n=1}^{\mathcal{N}/2} (|\uparrow_{2n}\rangle|\downarrow_{2n\pm1}\rangle - |\downarrow_{2n}\rangle|\uparrow_{2n\pm1}\rangle)/\sqrt{2} . \qquad (8.12)$$

We shall prove that

$$H^{MG}|d\rangle_\pm = 0, \qquad (8.13)$$

and that all other eigenenergies are positive. Thus, $|d\rangle_\pm$ span the ground state manifold of H^{MG}. H^{MG} includes antiferromagnetic interactions between next nearest neighbors, which partially frustrates the nearest neighbor correlations. Thus, we can expect the ground state to be more disordered than the pure nearest neighbor model. Indeed, the correlations of the Bethe wave function of the nearest neighbor model decay as an inverse power of distance, while by (8.9), the dimer state correlations vanish beyond a single lattice constant.

The proof of (8.13) is instructive since it demonstrates a general technique in constructing parent Hamiltonians. The total spin of a triad of spins at sites $(i-1, i, i+1)$ is

$$\mathbf{J}_i = \mathbf{S}_{i-1} + \mathbf{S}_i + \mathbf{S}_{i+1}, \tag{8.14}$$

whose square has eigenvalues $J(J+1)$, where $J = \frac{1}{2}, \frac{3}{2}$. The basic idea is to express H^{MG} as a sum of projection operators:

$$H^{MG} = |K| \sum_i \mathcal{P}_{3/2}(i-1, i, i+1), \tag{8.15}$$

where

$$\begin{aligned}
\mathcal{P}_{3/2}(i-1, i, i+1) &= \frac{1}{3}\left(\mathbf{J}_i^2 - \frac{3}{4}\right) \\
&= \frac{1}{2} + \frac{2}{3}\left(\mathbf{S}_{i-1} \cdot \mathbf{S}_i + \mathbf{S}_{i-1} \cdot \mathbf{S}_{i+1} + \mathbf{S}_i \cdot \mathbf{S}_{i+1}\right).
\end{aligned} \tag{8.16}$$

Clearly, $\mathcal{P}_{3/2}$ annihilates any state with total spin $J = \frac{1}{2}$ of the triad $(i-1, i, i+1)$. Also, dimer states (8.12) do not contain states with total $J^z > \frac{1}{2}$ of any three sites, since two of the three spins have cancelling S^z quantum numbers.

This implies that there are no triads with total spin $J > \frac{1}{2}$, due to the following argument: assume a triad with $J^z = \frac{1}{2}$ but $J = \frac{3}{2}$. Now perform a global rotation on $|d\rangle$. This will admix (by applications of $\mathbf{J}_i^+$) a component of $J_i^z = \frac{3}{2}$ into the wave function, which contradicts the rotational invariance of $|d\rangle$ (see discussion before (8.3)). Thus, there cannot be any $J > \frac{1}{2}$ component in $|d\rangle$, and therefore each operator $\mathcal{P}_{3/2}$ annihilates $|d\rangle_{\pm}$. Since $|K|\mathcal{P}_{3/2}$ are non-negative operators, $|d\rangle_{\pm}$ span the ground state manifold of H^{MG}. Q.E.D.

8.2.2 SQUARE LATTICE RVB STATES

On the square lattice of $S = \frac{1}{2}$, the resonating valence bond (RVB) states (8.3) fulfill all the requirements of Marshall's theorems 5.1 and 5.2 for the quantum antiferromagnet. These states were proposed as candidates for spin liquid ground states by P.W. Anderson. Although the nearest neighbor antiferromagnet on the square lattice is (by all indications) ordered at $T = 0$, these states have been very popular in the context of doped antiferromagnets and high-temperature superconductivity (see bibliography and Chapter 19).

Unfortunately, it is not easy to evaluate correlations of RVB states. One can see, by (8.10), that their components $|\alpha\rangle$ are not orthogonal. The number of different bond coverings on the square lattice increases exponentially with the number of sites. M.E. Fisher has calculated (see bibliography) that the number of dimer configurations on the square lattice of size $\mathcal{N}$ grows as

$$\text{Number of dimers} \sim (1.791623)^{\mathcal{N}/2}. \tag{8.17}$$

Numerical Monte Carlo simulations by Liang, Doucot, and Anderson on finite lattices suggest that the RVB state has no long-range order for bonds that decay at least as rapidly as

$$u_{ij} \propto |\mathbf{x}_i - \mathbf{x}_j|^{-p} , \qquad p \geq 5. \tag{8.18}$$

The RVB states can thus be used as variational ground states for both ordered and disordered phases. This makes them appealing candidates for studying the transition from the Néel antiferromagnet to a paramagnetic phase.

8.3   Valence Bond Solids and AKLT Models

The *valence bond solids* (VBS) are

$$|\Psi^{VBS}\rangle = \prod_{\langle ij \rangle} \left(a_i^\dagger b_j^\dagger - b_i^\dagger a_j^\dagger \right)^M |0\rangle, \tag{8.19}$$

where $\langle ij \rangle$ are all the nearest neighbor bonds of the lattice and M is an integer which obeys

$$M = 2S/z . \tag{8.20}$$

z is the lattice coordination number. It is clear that (8.20) restricts S to values depending on the lattice structure. For instance, for the one-dimensional lattice $S = 1, 2\dots$, while for the square lattice $S = 2, 4\dots$. Some VBS states are depicted in Fig. 8.4.

Affleck, Kennedy, Lieb, and Tasaki (AKLT) have constructed the parent Hamiltonians for the VBS states as follows:

$$\mathcal{H}^{AKLT} = \sum_{\langle ij \rangle} \sum_{J=2S-M+1}^{2S} K_J \mathcal{P}_J(ij) , \quad K_J \geq 0 . \tag{8.21}$$

The bond projector $\mathcal{P}_J(ij)$ projects the *bond spin* $\mathbf{J}_{ij} = \mathbf{S}_i + \mathbf{S}_j$ onto the subspace of magnitude J. Any power of $m = 0, 1, 2, \dots$ of $\mathbf{S}_i \cdot \mathbf{S}_j$ can be written in terms of powers of $\mathbf{J}_{ij}^2$ and expanded as a linear combination of bond projection operators

$$(\mathbf{S}_i \cdot \mathbf{S}_j)^m = \sum_{J=0}^{2S} [\frac{1}{2}J(J+1) - S(S+1)]^m \mathcal{P}_J(ij) . \tag{8.22}$$

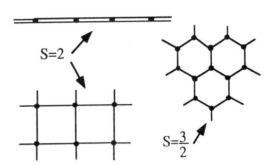

FIGURE 8.4. Some valence bond solids.

Conversely, (8.22) can be inverted to express the bond projection operators as polynomials of $\mathbf{S}_i \cdot \mathbf{S}_j$.

To prove that (8.19) is the ground state of (8.21), we show that

$$\mathcal{H}^{AKLT} |\Psi^{VBS}\rangle = 0, \qquad (8.23)$$

which holds if $|\Psi^{VBS}\rangle$ has no component with bond spin $J(ij) > 2S - M$, for any (ij).

Let us consider the contribution to $|\Psi^{VBS}\rangle$ from the term with the maximally possible number of $a^\dagger$'s on a particular bond (ij):

$$\cdots (a_i^\dagger)^{2S-M} \left(a_i^\dagger b_j^\dagger - b_i^\dagger a_j^\dagger \right)^M (a_j^\dagger)^{2S-M} \cdots . \qquad (8.24)$$

By counting the power of $a^\dagger$'s minus $b^\dagger$'s on the bond, we find that the maximal eigenvalue of J_{ij}^z in (8.19) is

$$J_{max}^z = 2S - M. \qquad (8.25)$$

If $|\Psi^{VBS}\rangle$ had a component with bond spin $J > J_{max}^z$, a global rotation of the full wave function would produce a component with $J^{z'} = J$. But $|\Psi^{VBS}\rangle$ is invariant under global rotations, and therefore (8.25) would be contradicted. Thus, $J_{max} = 2S - M$, and we have proven that $|\Psi^{VBS}\rangle$ is annihilated by $\mathcal{H}^{AKLT}$, which contains only projection operators with $J > J_{max}$. Since K_J in (8.21) are non-negative, the eigenvalues of $\mathcal{H}^{AKLT}$ are non-negative and $|\Psi^{VBS}\rangle$ is a ground state. Q.E.D.

An important example is the case of the spin one chain, where $S = 1$ and $M = 1$. By (8.22), the explicit form of the Hamiltonian reads

$$\mathcal{H}^{AKLT} = K \sum_{\langle ij \rangle} P_2(ij)$$

$$= K \sum_{\langle ij \rangle} \left[\mathbf{S}_i \cdot \mathbf{S}_j + \frac{1}{3} (\mathbf{S}_i \cdot \mathbf{S}_j)^2 + \frac{2}{3} \right]. \qquad (8.26)$$

This model adds a biquadratic term to the standard Heisenberg model. We shall see that the ground state and excitations of this model resemble those predicted by other approaches for the standard Heisenberg model. In that respect, one may conclude that the biquadratic perturbation is effectively "small."

8.3.1 CORRELATIONS IN VALENCE BOND SOLIDS

In order to calculate the correlations of the VBS states (8.19), we use the spin coherent states $|\hat{\Omega}\rangle$, defined in (7.18). This basis allows us to express the VBS correlations as a classical statistical mechanics average.

The overlap of a VBS state with a spin coherent state is

$$
\begin{aligned}
\Psi^{VBS}[\hat{\Omega}] &= \langle 0 | \prod_{\langle ij \rangle} (a_i b_j - b_i a_j)^M \prod_i \frac{(u_i a_i^\dagger + v_i b_i^\dagger)^{2S}}{\sqrt{(2S)!}} |0\rangle \\
&= \sqrt{(2S)!} \prod_{\langle ij \rangle} (u_i v_j - v_i u_j)^M \\
&= \sqrt{(2S)!} \prod_{\langle ij \rangle} \left(\frac{1 - \hat{\Omega}_i \cdot \hat{\Omega}_j}{2} \right)^{M/2}, \qquad (8.27)
\end{aligned}
$$

where $u(\theta, \phi), v(\theta, \phi)$ were defined in (7.14). The explicit expression for the VBS wave function in terms of unit vectors is very useful for understanding its correlations. By (7.26), the spin correlations are given by

$$
\begin{aligned}
\langle \mathbf{S}_i \cdot \mathbf{S}_j \rangle &= Z^{-1} (S + 1 - \delta_{ij})(S + 1) \\
&\quad \times \int \prod_i d\hat{\Omega}_i \left| \Psi^{VBS}[\hat{\Omega}] \right|^2 \hat{\Omega}_i \cdot \hat{\Omega}_j, \qquad (8.28)
\end{aligned}
$$

where the normalization denominator is

$$ Z = \int \prod_i d\hat{\Omega}_i \left| \Psi^{VBS}[\hat{\Omega}] \right|^2. \qquad (8.29) $$

An explicit calculation of (8.28) on an open one-dimensional lattice is possible by an iterative procedure.[4] For a distance of n lattice spacings, the result is

$$
\begin{aligned}
\langle \mathbf{S}_0 \cdot \mathbf{S}_n \rangle &= \begin{cases} (-1)^n (S+1)^2 \exp[-\kappa |n|] & n \neq 0 \\ S(S+1) & n = 0 \end{cases}, \\
\kappa(S) &= \ln \left(1 + \frac{2}{S} \right). \qquad (8.30)
\end{aligned}
$$

---

[4]See Arovas, Auerbach, and Haldane in the bibliography.

The magnitude of the correlations decays in a purely exponential fashion. We shall see in subsequent chapters that this behavior is similar to the prediction of Haldane's continuum approximation for the Heisenberg antiferromagnet, reviewed in Chapters 12 and 13. This supports our previous contention that (8.26) is not "far" from the standard Heisenberg model.

For higher dimensions one can understand the correlations by considering an equivalent classical statistical mechanics system. We define the *classical Boltzmann weight* as

$$\exp\left(-\frac{\Phi}{T}\right) \Leftrightarrow |\Psi^{VBS}[\hat{\Omega}]|^2 \; , \tag{8.31}$$

where the effective "temperature" is

$$T \Leftrightarrow \frac{1}{M} \; , \tag{8.32}$$

The classical "energy" is given by expanding near the Néel correlations, i.e., for small values of $(1 + \hat{\Omega}_i \cdot \hat{\Omega}_j)$,

$$\begin{aligned}
\Phi &= -\sum_{\langle ij \rangle} \ln[(1 - \hat{\Omega}_i \cdot \hat{\Omega}_j)/2] \\
&\approx \sum_{\langle ij \rangle} \left[\frac{1}{2} + \frac{1}{2}(\hat{\Omega}_i \cdot \hat{\Omega}_j) + \mathcal{O}(1 + \hat{\Omega}_i \cdot \hat{\Omega}_j)^2 \dots \right].
\end{aligned} \tag{8.33}$$

Φ is a classical Heisenberg-like Hamiltonian with short-range antiferromagnetic interactions and $O(3)$ rotational symmetry. By the classical version of Mermin and Wagner's Theorem 6.2 (see (6.35)), we expect no long-range order for finite T (or M) in one and two dimensions. Thus, the VBS states on one- and two-dimensional lattices have only short-range order, i.e., they describe "*quantum spin liquids.*" On the other hand, on three-dimensional lattices, for large enough M (i.e., low "temperatures"), we expect the classical Hamiltonian to produce long-range antiferromagnetic order. In that case, the VBS state describes a rotationally invariant state but with true antiferromagnetic long-range order.[5]

8.4 Exercises

1. Draw the valence bond states for the linear chain, the square, the honeycomb, the triangular, and the cubic lattices. What are the allowed values of S for each lattice?

2. For bipartite lattices, show that the valence bond solids (8.19) satisfy Marshall's sign criterion (5.13).

---

[5] A similar classical analogy trick was used by Laughlin for the fractional quantum Hall effect; see bibliography.

3. Prove the identity

$$\mathcal{P}_2(ij) = \mathbf{S}_i \cdot \mathbf{S}_j + \frac{1}{3}(\mathbf{S}_i \cdot \mathbf{S}_j)^2 + \frac{2}{3}. \tag{8.34}$$

4. The generating functional for correlations in the Schwinger boson mean field states is given by the functional

$$Z[j] = \langle \hat{u} | \exp \left[\sum_{im} a_{im}^\dagger j_{im} a_{im} \right] | \hat{u} \rangle. \tag{8.35}$$

Prove that

$$Z[\hat{j}] = \det_{im} \left| 1 - \hat{u} e^{\hat{j}} \hat{u} e^{\hat{j}} \right|^{-1/2}, \tag{8.36}$$

where

$$\hat{j}_{im,i'm'} \equiv \delta_{ii'} \delta_{mm'} j_{im}. \tag{8.37}$$

Hint: Insert resolutions of the identity using boson coherent states (see Appendix C) and express Z as a complex Gaussian integral.

5. Consider the one-dimensional mean field state $|\hat{u}\rangle$ where

$$\hat{u}_{ij} = u\,(\delta_{j,i+1} + \delta_{j,i-1}). \tag{8.38}$$

Use (8.36) with diagonal sources $j_{im} = j$, and write the average occupation equation as

$$n(u) = \sum_m \frac{\partial \ln Z[\hat{u}][}{\partial j_{im}} = 2S. \tag{8.39}$$

Solve for the function $u(S)$.

6. Using

$$S_i^z \equiv \frac{1}{2}(n_{i,\frac{1}{2}} - n_{i,-\frac{1}{2}}), \tag{8.40}$$

show by differentiating $Z(j)$ with respect to the appropriate source terms that the on-site mean field fluctuations are given by

$$\frac{\langle \hat{u}^{VBS} | (S_i^z)^2 | \hat{u}^{VBS} \rangle}{\langle \hat{u}^{VBS} | \hat{u}^{VBS} \rangle} = \frac{1}{2}S(S+1). \tag{8.41}$$

Using rotational invariance, relate the result to $\langle \mathbf{S}^2 \rangle$. What is the error due to the neglect of the local constraints?

7. Using $Z(j)$ of (8.36), calculate the mean field correlation function of the one-dimensional valence bond state:

$$\frac{\langle \hat{u}^{VBS} | S_0^z S_n^z | \hat{u}^{VBS} \rangle}{\langle \hat{u}^{VBS} | \hat{u}^{VBS} \rangle}, \tag{8.42}$$

and compare your result to the exact VBS correlations in (8.30).

Bibliography

Several ideas and examples in this chapter are taken from the lecture notes of

- D.P. Arovas and S.M. Girvin, *Exact Answers to Some Interesting Questions*, Proc. 7th International Conference on Many Body Theories, Minneapolis (1991).

The AKLT model was introduced in

- I. Affleck, T. Kennedy, E.H. Lieb, and H. Tasaki, Phys. Rev. Lett. **59**, 799 (1987).

The Majumdar–Ghosh model was defined in

- C.K. Majumdar and D.K. Ghosh, J. Math Phys. **10**, 1388 (1969).

The exact correlations of the valence bond solids in one dimension were evaluated by

- D.P. Arovas, A. Auerbach, and F.D.M. Haldane, Phys. Rev. Lett. **60**, 531 (1988).

Resonating valence bonds and their possible connection to high-temperature superconductivity was proposed by

- P.W. Anderson, Science **235**, 1196 (1987);
- S. Kivelson, D.Rokhsar, and J. Sethna, Phys. Rev. B **35**, 8865 (1987).

Equation (8.17) was derived by

- M.E. Fisher, Phys. Rev. **124**, 1664 (1961).

For calculations of correlations in RVB states, see

- W. Sutherland, Phys. Rev. B **37**, 3786 (1988);
- M. Kohmoto and Y. Shapir, Phys. Rev. Lett. **61**, 9439 (1988);
- S. Liang, B. Doucot, and P.W. Anderson, Phys. Rev. Lett. **61**, 365 (1988);
- E. Fradkin, *Field Theories of Condensed Matter Systems* (Addison-Wesley, 1991), Chapter 6.

A $1/N$ expansion for the correlations of valence bond states was carried out in

- M. Raykin and A. Auerbach, Phys. Rev. B **47**, 5118 (1993).

Unexpected agreement of low-order results with the exact correlations (8.30) were found.

An application of the projector method to the Laughlin wave function of the fractional quantum Hall effect is reviewed in

- F.D.M. Haldane, *The Quantum Hall Effect*, edited by R.E. Prange and S.M. Girvin (Springer-Verlag, 1987), Chapter 8.

The famous "Classical Plasma Analogy," which allows calculation of correlations in the Laughlin state of the fractional quantum Hall effect, is very similar in spirit to (8.31). See

- R.B. Laughlin, Phys. Rev. Lett. **50**, 1395 (1983);
- Chapter 7 of the book *The Quantum Hall Effect*, listed above.

9

From Ground States to Excitations

The focus of previous chapters has been primarily on the ground states of the Heisenberg model. Knowledge of the exact ground state—or in lack of one, of a good variational state—enables us to calculate the equal time spin correlations at $T = 0$. Experiments, however, can measure *dynamical responses* at finite frequencies and at finite temperatures. It is clear from linear response theory (see Appendix B) that dynamical correlations depend on the *excited* states and energies.

In this chapter, we shall use the ground state correlations to construct certain approximate low-lying excitations. This approach is called the *single mode approximation* (SMA).[1]

The Heisenberg antiferromagnet provides an opportunity to demonstrate the SMA for a system with a highly correlated ground state.

The information about spin excitations and correlations is embodied in the dynamical structure factor (B.15):

$$
\begin{aligned}
S(\mathbf{q}, \omega) &= \mathcal{N}^{-1} \int_{-\infty}^{\infty} dt\, e^{i\omega t} \sum_{ij} e^{i\mathbf{q}(\mathbf{x}_i - \mathbf{x}_j)} \langle S_i^z(t) S_j^z(0) \rangle \\
&= \frac{2\pi}{\mathcal{N} Z} \sum_{\alpha, \beta} e^{-E_\alpha/T} \langle \alpha | S_{\mathbf{q}}^z | \beta \rangle \langle \beta | S_{-\mathbf{q}}^z | \alpha \rangle\, \delta(\omega + E_\alpha - E_\beta) \,.
\end{aligned}
$$

$$(9.1)$$

The equal-time correlation function is

$$
\begin{aligned}
S(\mathbf{q}) &= \int_{-\infty}^{+\infty} \frac{d\omega}{2\pi}\, S(\mathbf{q}, \omega) \\
&= \mathcal{N}^{-1} \langle S_{\mathbf{q}}^z S_{-\mathbf{q}}^z \rangle,
\end{aligned}
$$

$$(9.2)$$

which depends only on the ground state wave function. Another useful function is the "double-commutator" correlation function, previously encountered in the proof of Mermin and Wagner's theorem (see (6.22)):

$$
F(\mathbf{q}) = \mathcal{N}^{-1} \left\langle \left[S_{-\mathbf{q}}^z, [\mathcal{H}, S_{\mathbf{q}}^z] \right] \right\rangle
$$

---

[1] The SMA was originally used by Bijl and Feynman to determine the phonon-roton dispersion curve in superfluid ^{4}He.

$$= \int_{-\infty}^{+\infty} \frac{d\omega}{2\pi} \omega \, S(\mathbf{q}, \omega). \tag{9.3}$$

(The different choice of spin direction here and in (6.22) is unimportant for a rotationally invariant Hamiltonian.) Although $F(\mathbf{q})$ is a static ground state expectation value, it contains dynamical information through $\mathcal{H}$.

Equations (9.2) and (9.3) show that $S(\mathbf{q})$ and $F(\mathbf{q})$ describe the average and the first frequency moment of $S(\mathbf{q}, \omega)$, respectively. Thus, a characteristic frequency for spin excitations at momentum $\mathbf{q}$ is defined as

$$\bar{\omega}_{\mathbf{q}} \equiv \frac{F(\mathbf{q})}{S(\mathbf{q})}. \tag{9.4}$$

9.1   The Single Mode Approximation

We denote the ground state by $|0\rangle$. By translational invariance, all excitations can be labelled by the lattice momentum $\mathbf{q}$. A *"single mode"* state is constructed by

$$|\mathbf{q}\rangle = S_{\mathbf{q}}^z |0\rangle. \tag{9.5}$$

The single mode state approximates a true excitation of the system if $S(\mathbf{q}, \omega)$ at $T = 0$, is sharply peaked about $\omega = \bar{\omega}_{\mathbf{q}}$, i.e.,

$$S(\mathbf{q}, \omega) \approx 2\pi S(\mathbf{q})\delta(\omega - \bar{\omega}_{\mathbf{q}}) . \tag{9.6}$$

By the second line in (9.1), (9.6) will be precise if

$$(\mathcal{H} - E_0)|\mathbf{q}\rangle = \bar{\omega}_{\mathbf{q}}|\mathbf{q}\rangle. \tag{9.7}$$

In general, even if (9.7) holds exactly, the single mode states are only a small subset of all excitations—those connected to the ground state via $S_{\mathbf{q}}^z$. The number of single mode states increases with the size of the system. If their products are also approximate excitations, they can be considered as weakly interacting *elementary excitations*. Thus, their energies approximately determine the low-frequency response and low-temperature thermodynamics of the model.

In Section 6.3, we saw that a gap Δ between the ground state, and the lowest excited state which does not vanish in the thermodynamic limit, implies absence of long-range order at $T = 0$. In this case, the structure factor, by (9.1), vanishes below the gap frequency:

$$S(\mathbf{q}, \omega) = 0 , \qquad \omega \in (0, \Delta) , \tag{9.8}$$

and by (9.2) – (9.4), the single-mode energy $\bar{\omega}_{\mathbf{q}}$ is an *upper bound* on the lowest excitation energy at momentum $\mathbf{q}$:

$$\min_{\alpha} E_{\alpha}(\mathbf{q}) \leq \bar{\omega}_{\mathbf{q}}. \tag{9.9}$$

Unfortunately, even if there is a gap Δ^{SMA} in the single-mode spectrum

$$\Delta^{SMA} \leq \bar{\omega}_{\mathbf{q}} , \quad \forall \mathbf{q}. \tag{9.10}$$

This does not imply a true gap for the exact eigenenergies E_α.

9.2 Goldstone Modes

Inequality (9.9) cannot prove the existence of a gap, but it can prove gaplessness. Here we shall use it to prove the lattice version of Goldstone's theorem for the Heisenberg model. The essence of Goldstone's theorem is that for a Hamiltonian with short-range interactions (and no gauge fields) spontaneously broken continuous symmetry implies the existence of low-energy excitations called *Goldstone modes*. If the ground state has momentum $\bar{\mathbf{q}}$, the energy of the Goldstone mode vanishes as $\mathbf{q} \to \bar{\mathbf{q}}$. For example, $\bar{\mathbf{q}} = 0$ for the ferromagnet, and $\bar{\mathbf{q}} = \vec{\pi}$ for the Néel antiferromagnet in the cubic lattice. In relativistic field theories, Goldstone modes represent massless particles. In condensed matter physics, Goldstone modes appear in many systems: e.g., acoustic phonons in solids that break translational symmetry, spin waves in $O(n)$ Heisenberg models, $n > 1$, where n is the dimension of the spins.

The proof of the Goldstone theorem is given in detail by Lange (see bibliography). Here we shall present a shorter version,[2] which makes use of the SMA bound (9.9).

We consider the short-range Heisenberg Hamiltonian which satisfies the conditions for the Mermin and Wagner theorem (see Section 6.2):

$$\mathcal{H} = \frac{1}{2} \sum_{ij} J_{ij} \mathbf{S}_i \cdot \mathbf{S}_j, \tag{9.11}$$

where

$$\bar{J} = \frac{1}{2N} \sum_{ij} |J_{ij}| \, |\mathbf{x}_i - \mathbf{x}_j|^2 \; < \infty. \tag{9.12}$$

Goldstone's Theorem 9.1 *If the spin correlation diverges at some wave vector $\bar{\mathbf{q}}$:*

$$\lim_{\mathbf{q} \to \bar{\mathbf{q}}} S(\bar{\mathbf{q}}) \; \to \; \infty, \tag{9.13}$$

then there exists a Goldstone mode labelled by the momentum $\mathbf{q}$, whose energy $E(\mathbf{q})$ vanishes at $\bar{\mathbf{q}}$:

$$\lim_{\mathbf{q} \to \bar{\mathbf{q}}} E(\mathbf{q}) \; = 0. \tag{9.14}$$

---

[2] The author thanks Daniel Arovas for this proof.

We use the bound on $F(\mathbf{q})$ given by (6.26) for $h = 0$:

$$F(\mathbf{q}) \leq 2S(S+1)\bar{J}\,|\mathbf{q}|^2, \qquad (9.15)$$

and using (9.4) and (9.15) we find that

$$\bar{\omega}_{\bar{\mathbf{q}}} \leq \frac{2S(S+1)\bar{J}\,|\bar{\mathbf{q}}|^2}{S(\mathbf{q})}. \qquad (9.16)$$

Thus, by (9.9) and (9.13), Eq. (9.14) is proven. Q.E.D.

Corollary 9.2 *If the ground state has broken symmetry, there exist Goldstone modes as in Theorem 9.2.*

This follows directly from (6.6), which shows that broken symmetry imples true long-range order and the divergence of $S(\mathbf{q})$ as $\mathcal{N} \to \infty$. Q.E.D.

It is important to note that the converse of Goldstone's theorem is false, i.e., existence of gapless excitations does not imply true long-range order. Notable counterexamples are the Fermi gas which has gapless particle–hole excitations but no long-range order.[3] Also, the $S = \frac{1}{2}$ Heisenberg antiferromagnet in one dimension has gapless excitations which vanish at $q = 0, \pi$:

$$\omega_q \propto |\sin q| \ . \qquad (9.17)$$

However, its correlations decay algebraically at long distances.

9.3   The Haldane Gap and the SMA

In Section 8.3, the ground states of the AKLT models have been shown to be valence bond solids. The one-dimensional model for $S = 1$ is

$$\mathcal{H}^{AKLT} = K \sum_{\langle ij \rangle} \left[\mathbf{S}_i \cdot \mathbf{S}_j + \frac{1}{3}(\mathbf{S}_i \cdot \mathbf{S}_j)^2 + \frac{2}{3} \right]. \qquad (9.18)$$

The correlation function for the $S = 1$ VBS states was given in (8.30). Taking its Fourier transform, we obtain

$$S(q) = \frac{2(1 - \cos q)}{(5 + 3\cos q)}. \qquad (9.19)$$

$F(q)$ can also be evaluated for the valence bond wave function. The result[4] is

$$F(q) = \frac{10K}{27}(1 - \cos q). \qquad (9.20)$$

---

[3]Fermi liquids (when they exist) are also gapless, with no broken symmetry.
[4]See Arovas, Auerbach, and Haldane, bibliography of Chapter 8.

By (9.4), we find that the single-mode energies are

$$\bar\omega_{\mathbf{q}} = \frac{5K}{27}(5 + 3\cos q) \geq 0.370K. \tag{9.21}$$

As discussed earlier, although the single-mode spectrum has a gap of magnitude $0.370K$, this does not imply that a gap actually exists in the exact spectrum. However, numerical simulations for the AKLT model of $S = 1$ support the existence of a gap of magnitude $\Delta = 0.350K$.[5] The same simulation for the standard Heisenberg model is found to possess a gap of size $\Delta = 0.325K$. Since the AKLT model has a gap in the SMA spectrum, it suggests that the additional biquadratic terms $\frac{K}{3}(\mathbf{S}_i \cdot \mathbf{S}_j)^2$ do not drastically alter the ground state correlations and low excitations of the $S = 1$ Heisenberg model in one dimension.[6]

The gap, which is commonly called *Haldane's gap*, survives in the thermodynamic limit ($\mathcal{N} \to \infty$). It is a general feature of integer spin quantum antiferromagnets in low dimensions, which is explained by the continuum theory of Haldane in Chapters 12 and 15.

Bibliography

The single-mode approximation for superfluid ^{4}He is described in

- R.P. Feynman, *Statistical Mechanics* (Benjamin, New York, 1972), Chapter 11.

The SMA was applied to derive the magnetophonon and magnetoroton excitations in the fractional quantum Hall system by

- S.M. Girvin, A.H. Macdonald, and P.M. Platzman, Phys. Rev. B **33**, 2481 (1986).

The Goldstone theorem for the ferromagnet was proven by

- R.V. Lange, Phys. Rev. **146**, 301 (1966).

The following lecture notes on spontaneous symmetry breaking and Goldstone modes are recommended

- S. Coleman, *Aspects of Symmetry*, (Cambridge, 1985), Chapter 5.

---

[5]These are unpublished results of F.D.M. Haldane.

[6]See Arovas and Girvin (bibliography of Chapter 8) for the effects of biquadratic perturbations on the correlations.

Part III

Path Integral Approximations

Illustration by Dick Codor.

10

The Spin Path Integral

It is always hard to define what is meant by *"understanding"* a particular system. Even an exact analytical form for the ground state wave function and energy may not *explain* their physical properties. For instance, the Bethe ansatz wave functions for the $S = \frac{1}{2}$ Heisenberg model require a numerical computation of their correlations. Understanding a particular model sometimes means having a simple approximation to its properties. An approximation, though imprecise or perhaps plagued with mathematical ambiguities, can be more illuminating than the exact solution. In particular, an approximation can unify a family of models and treat an open neighborhood of an exactly soluble point in parameter space.

Path integrals provide formal expressions which lead to useful approximation schemes. Usually, they cannot be evaluated in closed analytic form.[1] Nevertheless, in spite of their shortcomings, they are valuable tools in theoretical physics. By providing compact expressions and notations, path integrals are often used to generate asymptotic expansions and to formulate mean field theories.

The spin coherent states path integral describes quantum spins in terms of time-dependent histories of unit vectors. This picture is highly intuitive since it connects the classical description and quantum phenomena. As such, it is a natural starting point for semiclassical approximations, which will be described in Chapters 11, 12, and 19.

10.1 Construction of the Path Integral

Spin coherent states can be used to construct a path integral representation of the Heisenberg model. The generating functional in the imaginary time formulation is[2]

$$Z[j] \;\; = \;\; \operatorname{Tr} T_\tau \left(\exp \left[- \int_0^\beta d\tau \mathcal{H}(\tau) \right] \right)$$

---

[1]In the words of A.M. Polyakov: *"There are no tables for path integrals."*
[2]See (B.17) in Appendix B.

$$= \lim_{N_\epsilon \to \infty} \mathrm{Tr}\, T_\tau \prod_{n=0}^{N_\epsilon - 1} [1 - \epsilon \mathcal{H}(\tau_n)] \; , \tag{10.1}$$

where β is the inverse temperature, T_τ is the time ordering operator, $\epsilon = \beta/N_\epsilon$ is the timestep, and $\tau_n = n\epsilon$ is the discrete imaginary time. The generating Hamiltonian includes source currents

$$\mathcal{H}(\tau) = \mathcal{H} - \sum_{i\alpha} j_i^\alpha(\tau) S_i^\alpha \; , \quad \tau \in [0, \beta) \; . \tag{10.2}$$

The basic ingredient in the construction of the path integral is the resolution of the identity provided by (7.25). Using (7.27), and inserting N_ϵ resolutions of the identity between the factors in (10.1) we obtain the multidimensional integral

$$Z[j] = \lim_{N_\epsilon \to \infty} \int \left(\prod_{\tau,i} d\hat{\Omega}_{i\tau} \right) \prod_{\tau=\epsilon}^{\beta} \langle \hat{\Omega}(\tau) | \hat{\Omega}(\tau - \epsilon) \rangle \, [1 - \epsilon H(\tau)] \; , \tag{10.3}$$

where $\hat{\Omega} = \left(\hat{\Omega}_1, \ldots, \hat{\Omega}_{\mathcal{N}} \right)$, and the "classical Hamiltonian" is defined as

$$H(\tau) \equiv \frac{\langle \hat{\Omega}(\tau) | \mathcal{H}(\tau) | \hat{\Omega}(\tau - \epsilon) \rangle}{\langle \hat{\Omega}(\tau) | \hat{\Omega}(\tau - \epsilon) \rangle}. \tag{10.4}$$

We define

$$\hat{\Omega}(\beta) = \hat{\Omega}(0) \tag{10.5}$$

and note that the measure includes only one integration for both times.

A priori, $\hat{\Omega}(\tau)$ is an arbitrary discrete function of τ. In the following, we shall do a blatantly illegal manipulation. We shall treat it as a continuous and differentiable function in the limit of $N_\epsilon \to \infty$ by substituting its differences by derivatives:

$$\frac{\hat{\Omega}_i(\tau + \epsilon) - \hat{\Omega}_i(\tau)}{\epsilon} \Rightarrow \dot{\hat{\Omega}}_i(\tau) + \mathcal{O}(\epsilon). \tag{10.6}$$

The implicit assumption in (10.6) is that Z is dominated by paths that are smooth, i.e., $|\dot{\hat{\Omega}}| < \infty$. This turns out to be *unjustified*.[3] By ignoring discontinuous paths, we lose information about ordering of operators in the quantum Hamiltonian. Nonetheless, we shall proceed to manipulate the "time derivatives" using the rules of differentiable functions. We must keep in mind that we are on shaky mathematical grounds. For that reason, path integral results should be checked whenever possible against operator methods which do not suffer from ordering ambiguities.

---

[3] See Klauder, and Chapter 31 in Shulman's book.

Using (7.19), we write and expand the overlap between coherent states at nearby timesteps to leading order in ϵ as follows:

$$\langle \hat{\Omega}(\tau + \epsilon) | \hat{\Omega}(\tau) \rangle \;\; = \;\; \exp\left(-iS\epsilon \sum_i \dot{\phi}_i \cos[\theta_i(\tau)] + \dot{\chi}_i \right), \quad (10.7)$$

where $\chi(\tau)$ is an arbitrary gauge convention. The classical Hamiltonian (10.4) multiplies ϵ, and can be evaluated at equal times:

$$H[\hat{\Omega}(\tau)] \rightarrow \langle \hat{\Omega}(\tau) | \mathcal{H}(\tau) | \hat{\Omega}(\tau) \rangle. \quad (10.8)$$

The limit $N_\epsilon \rightarrow \infty$ will transform (10.3) into a path integral. The path integration measure is defined as

$$\mathcal{D}\hat{\Omega}(\tau) \;\; = \;\; \lim_{N_\epsilon \rightarrow \infty} \prod_{i,n} d\hat{\Omega}_i(\tau_n). \quad (10.9)$$

By exponentiating the Hamiltonian, and discarding higher-order terms in ϵ, we take the *formal* continuum limit of (10.3) and write

$$Z[j] \;\; = \;\; \oint \mathcal{D}\hat{\Omega}(\tau) \, \exp\left(-\tilde{S}[\hat{\Omega}] \right),$$

$$\tilde{S}[\hat{\Omega}] \;\; = \;\; -iS \sum_i \omega[\hat{\Omega}_i] + \int_0^\beta d\tau \, H[\hat{\Omega}(\tau)]. \quad (10.10)$$

The notation $\oint$ reflects the periodic boundary condition (10.5). For the gauge convention $\chi_i(\tau) = 0$, the functional ω depends on the history of a single spin as follows:

$$\omega[\hat{\Omega}] \;\; = \;\; -\int_0^\beta d\tau \, \dot{\phi} \cos\theta$$

$$= \;\; -\int_{\phi_0}^{\phi_0} d\phi \, \cos(\theta_\phi). \quad (10.11)$$

We see that ω is geometric: it depends on the *trajectory* of $\hat{\Omega}_i(\tau)$ on the sphere and not on its explicit time dependence. The functional $S\omega$ is also called the *Berry phase* of the spin history, since it describes the phase acquired by a spin that aligns with an adiabatically rotating external magnetic field[4] which is parallel to $\hat{\Omega}(\tau)$.

Now we shall see that the Berry phase measures the area enclosed by the path $\hat{\Omega}(\tau)$ on the unit sphere. A path increment $d\hat{\Omega}$ is connected to the north pole (as shown in Fig. 10.1) by the longitudes ϕ and $\phi + d\phi$. This area increment is a triangle on the sphere whose area is given by

$$d\omega' \;\; = \;\; (1 - \cos\theta)d\phi. \quad (10.12)$$

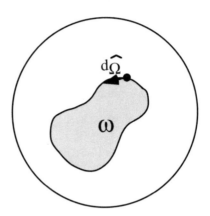

FIGURE 10.1. A spin history as an orbit on the unit sphere.

Thus, the area enclosed by the closed orbit parametrized by $\theta(\tau), \phi(\tau)$ is given by

$$\omega' = \int_0^\beta d\tau \dot{\phi}[1 - \cos\theta(\tau)]. \tag{10.13}$$

ω' is therefore the area enclosed on the left of the counterclockwise orbit $\{\hat{\Omega}(\tau)\}_0^\beta$. The identity

$$\omega' = \omega \tag{10.14}$$

holds if $\phi(\beta)$ does not cross the "date line" boundary $\pm\pi$, as required by the convention (7.17).[5]

It is useful to express the Berry phase ω in a gauge invariant form, that is to say without specifying any parametrization of the sphere such as θ, ϕ. We introduce a vector potential $\mathbf{A}(\hat{\Omega})$ which satisfies

$$\omega = \int_0^\beta d\tau \, \mathbf{A}(\hat{\Omega})\dot{\hat{\Omega}}. \tag{10.15}$$

$\mathbf{A}(\hat{\Omega})$ is a unit magnetic monopole vector potential whose line integral over the orbit $\{\hat{\Omega}(\tau)\}$ is equal to the solid angle ω subtended by that orbit. By Stokes theorem, $\mathbf{A}$ satisfies

$$\nabla \times \mathbf{A} \cdot \hat{\Omega} = \epsilon^{\alpha\beta\gamma} \frac{\partial A^\beta}{\partial \hat{\Omega}^\alpha} \hat{\Omega}^\gamma = 1, \tag{10.16}$$

---

[4]See the Exercises, and Shapere and Wilczek.

[5]See Exercise 2 for a clarification of the difference between ω and ω'.

where $\epsilon^{\alpha\beta\gamma}$ is the fully antisymmetric tensor, $\alpha, \beta, \gamma \in \{x, y, z\}$, and summation over repeated indices is assumed. Two standard choices for $\mathbf{A}$ are

$$\mathbf{A}^a = -\frac{\cos\theta}{\sin\theta}\hat{\phi} , \tag{10.17}$$

$$\mathbf{A}^b = \frac{1 - \cos\theta}{\sin\theta}\hat{\phi}$$

$$= \frac{\hat{z} \times \hat{\Omega}}{\hat{z} \cdot \hat{\Omega} + 1}. \tag{10.18}$$

The gauge $\mathbf{A}^b$ differs from $\mathbf{A}^a$ by the location of its singularities. $\mathbf{A}^a$ is singular at the north and south poles. The domain of ϕ is $[-\pi, \pi)$, and thus paths that cross the "date line" at $\phi = \pm\pi$ are not allowed. On the other hand, $\mathbf{A}^b$ has only one singularity at the south pole $\hat{\Omega} = -\hat{z}$. This is where the "Dirac string," which carries the magnetic monopole's flux, enters the sphere. For an infinitesimal orbit around the south pole, the value of $S\omega$ is equal to $4\pi S$. Since S is an integer multiple of half, $\exp[-iS\omega]$ is a continuous functional of the orbit even as it crosses the singularity of the gauge field at the south pole.

10.1.1   THE GREEN'S FUNCTION

The *Green's function* $G(t)$ describes real-time evolution at zero temperature. It is given by the matrix element of the evolution operator between two coherent states:

$$G(\hat{\Omega}_0, \hat{\Omega}_t; t) = \langle\hat{\Omega}_t|T_{t'}\left(\exp\left[-i\int_0^t dt'\,\mathcal{H}(t')\right]\right)|\hat{\Omega}_0\rangle, \tag{10.19}$$

where $T_{t'}$ is the real-time ordering operator (see B.3). $G(t)$ can be represented by a path integral in close analogy to the path integral of Z. In the derivations of (10.1)–(10.10), one replaces the imaginary time variable τ by real time t,

$$\tau \rightarrow it' ,$$
$$t' \in [0, t], \tag{10.20}$$

and the interval $[0, t]$ is discretized into N_ϵ timesteps of width ϵ. After inserting resolutions of the identity between factors of $(1 - i\epsilon\mathcal{H})$, and taking the continuum limit $N_\epsilon \to \infty$ (which introduces the aforementioned mathematical ambiguities due to the time derivatives), we obtain the formal expression

$$G(t) = \int_{\hat{\Omega}_0}^{\hat{\Omega}_t} \mathcal{D}\hat{\Omega}(t')\,\exp\left[iS[\hat{\Omega}]\right] , \tag{10.21}$$

where S is the real-time action

$$S[\hat{\Omega}] = \int_0^t dt'\left(S\sum_i \mathbf{A}\cdot\dot{\hat{\Omega}} - H[\hat{\Omega}(t'), t']\right). \tag{10.22}$$

10.2   The Large S Expansion

The spin coherent states path integrals, (10.10) and (10.21), are convenient starting points for deriving the semiclassical approximation. The coordinates are unit vectors, i.e., classical spins. The quantum effects enter through their (real or imaginary) time dependence. We can scale the parameters of $H[\hat{\Omega}]$ to be independent of S, i.e., use the corresponding classical Hamiltonian $H^{cl}[\hat{\Omega}]^6$. Thus, by sending $S \to \infty$, all time-dependent paths with $\dot{\hat{\Omega}} \neq 0$ are suppressed by the rapid oscillations of the Berry phase factor. Thus, the classical partition function is recovered:

$$\lim_{S \to \infty} Z[j] \sim Z' \int D\hat{\Omega} \, \exp\left[-\beta H^{cl}[\hat{\Omega}]\right], \qquad (10.23)$$

where Z' includes normalization factors and higher-order quantum corrections. The integral on the right-hand side is the generating functional for the classical Hamiltonian.

It is also possible to use S as the control parameter for a systematic asymptotic expansion of the Green's function. This semiclassical expansion introduces the quantum corrections to the classical theory. The first step is to rescale the time variable

$$
\begin{aligned}
\tau &\to S\tau = \bar{\tau}, \\
\beta &\to S\beta = \bar{\beta}.
\end{aligned}
\qquad (10.24)
$$

The classical inverse temperature $\bar{\beta}$ is taken to be independent of S. This allows us to scale S out of the action

$$
\begin{aligned}
Z(\bar{\beta}) &= \oint D\hat{\Omega}(\bar{\tau}) \, \exp\left(-S \, S^{cl}\left[\hat{\Omega}(\bar{\tau}), \bar{\beta}\right]\right), \\
S^{cl} &= \int_0^{\bar{\beta}} d\bar{\tau} \left(i \sum_i \mathbf{A} \cdot \dot{\hat{\Omega}} + H^{cl}[\hat{\Omega}(\bar{\tau})]\right).
\end{aligned}
\qquad (10.25)
$$

Now we can apply the method of steepest descents to Z (see Appendix E), using S as the large parameter. We obtain a sum over saddle points $\hat{\Omega}^{cl,\alpha}$ (see E.14):

$$Z \sim \sum_\alpha \exp\left(-S \, S^{cl}[\hat{\Omega}^{cl,\alpha}, \bar{\beta}]\right) Z'_\alpha, \qquad (10.26)$$

where the saddle point equations are

$$\frac{\delta S^{cl}}{\delta \hat{\Omega}}\bigg|_{\hat{\Omega} = \hat{\Omega}^{cl,\alpha}} = 0 . \qquad (10.27)$$

---

[6]See, for example, (6.35) and (6.36).

The prefactors Z'_α contain subdominant corrections in powers of S^{-1}. These can be evaluated by expanding the fluctuation integrals

$$Z'_\alpha = \oint \mathcal{D}\delta\hat{\Omega}\exp\left[-S\left(\mathcal{S}^{cl}[\hat{\Omega}] - \mathcal{S}^{cl}[\hat{\Omega}^{cl,\alpha}]\right)\right],\tag{10.28}$$

where $\delta\hat{\Omega} = \hat{\Omega} - \hat{\Omega}^{cl,\alpha}$.

10.2.1 SEMICLASSICAL DYNAMICS

The large S expansion of the Green's function requires us to scale

$$t \to St = \bar{t},\tag{10.29}$$

which yields

$$G(\bar{t}) = \int_{\hat{\Omega}_0}^{\hat{\Omega}_{\bar{t}}} \mathcal{D}\hat{\Omega}(\bar{t}')\,\exp\left(iS\,\mathcal{S}^{cl}\left[\hat{\Omega}\right]\right),$$

$$\mathcal{S}^{cl} = \int_0^{\bar{t}} d\bar{t}'\left(\sum_i \mathbf{A}\cdot\dot{\hat{\Omega}} - H^{cl}[\hat{\Omega}(\bar{t}')]\right).\tag{10.30}$$

$G(t)$ is dominated by time-dependent paths $\hat{\Omega}_{cl}(t')$, which extremize the action. Henceforth we shall replace $\bar{t} \to t$. The method of steepest descents yields a sum over saddle points (see E.14):

$$G \sim \sum_\alpha \exp\left(iS\,S[\hat{\Omega}^{cl,\alpha},\bar{t}]\right)G'_\alpha,\tag{10.31}$$

where $\hat{\Omega}^{cl}(t')$ are determined by the saddle point equations

$$\left.\frac{\delta}{\delta\hat{\Omega}}S[\hat{\Omega}]\right|_{\hat{\Omega}^{cl,\alpha}} = 0,\tag{10.32}$$

which are subject to the boundary conditions

$$\hat{\Omega}^{cl,\alpha}(0) = \hat{\Omega}_0,$$
$$\hat{\Omega}^{cl,\alpha}(t) = \hat{\Omega}_t.\tag{10.33}$$

The variation of the Berry phase part of the action is given by

$$\delta\omega[\hat{\Omega}] = \int_0^t dt'\,\delta\left(\mathbf{A}\cdot\dot{\hat{\Omega}}\right)$$

$$= \int_0^t dt'\left[\frac{\partial A^\alpha}{\partial\hat{\Omega}^\beta}\delta\hat{\Omega}^\beta\dot{\hat{\Omega}}^\alpha + A^\alpha\frac{d}{dt}\delta\hat{\Omega}^\alpha\right]$$

$$+ \left[\frac{\partial A^\alpha}{\partial\hat{\Omega}^\beta}\dot{\hat{\Omega}}^\beta\delta\hat{\Omega}^\alpha - \frac{\partial A^\alpha}{\partial\hat{\Omega}^\beta}\dot{\hat{\Omega}}^\beta\delta\hat{\Omega}^\alpha\right]$$

$$
\begin{aligned}
&= \int_0^t dt' \frac{\partial A^\alpha}{\partial \hat{\Omega}^\beta} \epsilon^{\alpha\beta\gamma} (\dot{\hat{\Omega}} \times \delta\hat{\Omega})_\gamma + \int_0^t dt' \frac{d}{dt'} \left(\mathbf{A} \cdot \delta\hat{\Omega} \right) \\
&= \int_0^t dt' \, \dot{\hat{\Omega}} \cdot (\dot{\hat{\Omega}} \times \delta\hat{\Omega}).
\end{aligned}
\tag{10.34}
$$

The integral over the total derivative vanishes since the endpoints are fixed by (10.33). In (10.34), we have also used (10.16) and the constant length of $\hat{\Omega}$, which yields

$$
\begin{aligned}
\delta\hat{\Omega} \cdot \hat{\Omega} &= \dot{\hat{\Omega}} \cdot \hat{\Omega} = 0 , \\
\dot{\hat{\Omega}} \times \delta\hat{\Omega} &\parallel \hat{\Omega} .
\end{aligned}
\tag{10.35}
$$

Applying (10.34) to (10.32) yields the classical *Euler–Lagrange* equations of motion

$$
\begin{aligned}
\hat{\Omega}^{cl} \times \dot{\hat{\Omega}}^{cl} &= \frac{\partial H[\hat{\Omega}^{cl}]}{\partial \hat{\Omega}} \\
\Rightarrow \quad \dot{\hat{\Omega}}_i^{cl}(t') &= \hat{\Omega}^{cl}(t') \times \left. \frac{\partial H}{\partial \hat{\Omega}_i(t')} \right|_{\hat{\Omega}^{cl}} .
\end{aligned}
\tag{10.36}
$$

Equations (10.36) describe a system of classical rotators in the "fast top" limit, i.e., when the rotators' internal rotational energy is much larger than the typical inter-rotator interaction energy. The right-hand side of the second row is the torque applied to rotator $\hat{\Omega}_i$, which changes its direction but not its magnitude.[7]

Using (10.36) we can verify that the Hamiltonian is a constant of motion on the classical path

$$
\begin{aligned}
\frac{d}{dt} H[\hat{\Omega}^{cl}(t')] &= \sum_i \frac{\partial H}{\partial \hat{\Omega}_i(t')} \cdot \dot{\hat{\Omega}}_i^{cl}(t') \\
&= \sum_i \frac{\partial H}{\partial \hat{\Omega}_i(t')} \cdot \left[\hat{\Omega}_i^{cl}(t') \times \frac{\partial H}{\partial \hat{\Omega}_i(t')} \right] \\
&= 0,
\end{aligned}
\tag{10.37}
$$

which ensures conservation of energy $H[\hat{\Omega}^{cl}(t)] = H[\hat{\Omega}_0]$ along the classical trajectory.

At this point, we encounter a problem: the equations of motion (10.36) are first order in time, but the solution must satisfy two boundary conditions (10.33) at $t' = 0$ and $t' = t$. This is impossible for almost all boundary conditions (e.g., when $H[\hat{\Omega}_t] \neq H[\hat{\Omega}_0]$). Klauder has suggested a way to overcome this problem by including second-order transient terms

---

[7]See Goldstein's book for the classical mechanics of rotators.

of order ϵ in the classical equations of motion. This allows one to solve a second-order equation, with the fixed boundary conditions, calculate the classical action and at the end, take the limit $\epsilon \to 0$.

10.2.2 SEMICLASSICAL SPECTRUM

The energy dependent Green's function is defined as

$$G(\hat{\Omega}_0, \hat{\Omega}_t; E) = i \int_0^\infty dt \, G(\hat{\Omega}_0, \hat{\Omega}_t; t) e^{i(E+i0^+)t}. \qquad (10.38)$$

The spectral function $\Gamma(E)$ is given by

$$\Gamma(E) = \int d\hat{\Omega}_0 \, G(\hat{\Omega}_0, \hat{\Omega}_0; E)$$

$$= \sum_\alpha \frac{1}{E - E_\alpha + i0^+}. \qquad (10.39)$$

The poles of $\Gamma(E)$ are the eigenenergies $\{E_\alpha\}$ of the Hamiltonian.

Here we derive in a very sketchy manner the leading-order semiclassical approximation for E_α using classical periodic orbits. The original derivation of the semiclassical spectrum using the path integral was by Gutzwiller, and his formula is widely used in quantizing Hamiltonians with chaotic classical dynamics.[8]

The path integral representation of $\Gamma(E)$ has the additional $d\hat{\Omega}_0$ and dt integrations. The semiclassical approximation to $\Gamma(E)$ was derived by Gutzwiller. The classical paths have duration t^E, which is determined by the saddle point approximation for the dt integral

$$\frac{\partial S[\hat{\Omega}^{cl}]}{\partial t} \bigg|_{t=t^E} + E = 0. \qquad (10.40)$$

Since the Berry phase term ω is geometric it does not depend on t^E. Thus,

$$H^{cl}[\hat{\Omega}^{cl}] = E, \qquad (10.41)$$

and

$$S^{cl}(t^E) + Et^E = \sum_i \omega[\hat{\Omega}_i]. \qquad (10.42)$$

Consider a periodic orbit $\hat{\Omega}_i^{E,\alpha}$ of energy E which is traversed once. As seen in the exercises, the saddle point equation for the trace over $\hat{\Omega}_0$ implies that $\dot{\hat{\Omega}}(0) = \dot{\hat{\Omega}}(t^E)$. Therefore, Γ is given by summing over all repetitions

---

[8] An area of research called *"Quantum Chaos"*; see bibliography.

of $\hat{\Omega}_i^{E,\alpha}$:

$$\Gamma \sim \sum_\alpha \sum_n \exp\left(inS \sum_i \omega[\hat{\Omega}_i^{E,\alpha}] \right)$$

$$= \sum_\alpha \frac{\exp\left(iS \sum_i \omega[\hat{\Omega}_i^{E,\alpha}] \right)}{1 - \exp\left(iS \sum_i \omega[\hat{\Omega}_i^{E,\alpha}] \right)}. \tag{10.43}$$

We compare (10.39) and (10.43). The semiclassical spectrum can be obtained from the poles of (10.43), E_α, which are determined by the *Bohr–Sommerfeld quantization condition*

$$\sum_i \omega[\hat{\Omega}_i^{E,\alpha}] = 2n\pi/S \Rightarrow E_\alpha^{sc}. \tag{10.44}$$

We use the theorem that two functions are the same, upto a constant, if they share the same poles and residues. Thus, for energies E_α^{sc} that are of order one (i.e., $n = \mathcal{O}(S)$), (10.44) approximates the quantum spectrum by

$$E_\alpha = E_\alpha^{sc} + \mathcal{O}(1/S). \tag{10.45}$$

10.3   Exercises

1. The gauge convention for the Euler angle $\chi = 0$ in the definition (7.15) of the coherent states leads to the expression (10.11) for the Berry phase. For the choice of $\chi = -\phi$, show that ω is now given by (10.13).

2. (*Patrick Lee's query*): The angle ϕ is defined in (7.17) to be in the half closed interval $[-\pi, \pi)$. Show that if the orbit $\hat{\Omega}(t)$ is allowed to cross the "date line" $\phi = -\pi$, the two expressions for ω, (10.11) and (10.13) differ by 2π. This corresponds to an overall factor of $\exp(2\pi S) = -1$ for half-odd integer S. To resolve this difficulty, calculate the line integral of (10.11) which surrounds the date line as follows: At longitude $\phi = -\pi + \epsilon$ it goes from θ to the north pole, surrounds the north pole, and comes down to θ on the other side of the date line at $\phi = -\pi - \epsilon$.

3. Consider a slowly varying magnetic interaction

$$H[\hat{\Omega}(\tau), \tau] = -h(\tau)\hat{\Omega}(\tau) \cdot \mathbf{S}, \tag{10.46}$$

where $\tau \in [0, \beta]$ parametrizes the field's trajectory. Assuming that $h(\tau) > 0$ for all τ, we define the nondegenerate adiabatic ground state as $\Psi_0(\tau)$. Show that Berry's phase, as defined by

$$S\omega = \lim_{dh/d\tau \to 0} \int_0^\beta d\tau\, \mathrm{Im}\langle \frac{d}{d\tau}\Psi_0(\tau)|\Psi_0(\tau)\rangle, \tag{10.47}$$

is equal to the expression (10.11).

Bibliography

Spin coherent state path integrals are reviewed by

- J. Klauder, Phys. Rev. D **19**, 2349 (1979).

A recommended textbook on path integrals and the semiclassical approximation is

- L.S. Schulman, *Techniques and Applications of Path Integration* (Wiley, 1981).

On Berry's phase and its numerous appearances in different areas of physics, see

- A. Shapere and F. Wilczek, *Geometric Phases in Physics* (World Scientific, 1989).

Classical equations of motion for rotators are solved in

- H. Goldstein, *Classical Mechanics* (Addison-Wesley, 1981), Chapter 5.

The semiclassical approximation to the quantum energy levels was derived by

- M.C. Gutzwiller, J. Math. Phys. **12**, 343 (1971).

A recent review on this subject with application to quantum chaos is

- M.C. Gutzwiller, *Chaos and Quantum Physics*, edited by M.J. Giannoni, A. Voros, and J. Zinn-Justin (North Holland, 1992).

11

Spin Wave Theory

In Chapter 10, we constructed the spin path integral and demonstrated that its large-S limit recovers the thermodynamics and dynamics of classical spins. Here we shall specialize to the quantum Heisenberg model and derive the low-order (harmonic) spin wave theory. This approach allows a physically appealing treatment of the long-range ordered phases, since the quantum effects enter as (real or imaginary) time dependent fluctuations about the classical ground state. The spin wave modes and dispersions are obtained from the linearized classical equations of motion. To lowest order in $1/S$, spin waves are independent harmonic oscillators which can be quantized semiclassically. A great advantage of this approach is that it can be followed for any frustrated Heisenberg model with a complicated ground state. We shall later specialize to the simpler ferromagnetic and antiferromagnetic cases.

The path integral, however, suffers from the ambiguity in operator ordering due to the continuous definition of the time derivative in the kinetic term (10.6). We shall rederive spin wave theory using Holstein–Primakoff bosons, previously introduced in Section 11.2. This will settle the ambiguity in the quantum correction to the ground state energy.

11.1   Spin Waves: Path Integral Approach

We consider the quantum Heisenberg Hamiltonian

$$\mathcal{H} = \frac{1}{2} \sum_{i,j} J_{ij} \mathbf{S}_i \cdot \mathbf{S}_j. \tag{11.1}$$

Using (7.34) (in Exercise 2), the expectation value of $\mathcal{H}$ in the coherent state $|\hat{\Omega}\rangle$ is

$$H[\hat{\Omega}] = \langle \hat{\Omega} | \mathcal{H} | \hat{\Omega} \rangle = \frac{S^2}{2} \sum_{i,j} J_{ij} \hat{\Omega}_i \cdot \hat{\Omega}_j. \tag{11.2}$$

The classical ground states of H depend on the details of the coupling constants J_{ij}. By the $O(3)$ symmetry of the model, global spin rotations generate a continuous manifold of degenerate ground states. Certain models have additional ground state degeneracies due to *frustration*. Frustration occurs when no configuration can minimize all individual bond interactions

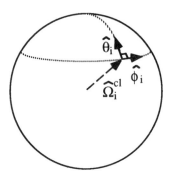

FIGURE 11.1. Parametrization of spin fluctuations about $\hat{\Omega}^{cl}$.

simultaneously. Degeneracies that are not related to symmetries of $\mathcal{H}$ are usually lifted by thermal and quantum fluctuations.[1] Here we shall not discuss frustrated cases.

The basic assumption of spin wave theory is that we can choose a member of the classical ground state manifold $\hat{\Omega}^{cl}$, and expand the partition function and Green's function about it in a saddle point expansion, controlled by the size of the spin S as explained in Section 10.2.

We choose the two transverse unit vectors at each site $\hat{\phi}_i, \hat{\theta}_i$, which describe the azimuthal and longitudinal directions at $\hat{\Omega}_i^{cl}$, as shown in Fig. 11.1:

$$\hat{\phi}_i \times \hat{\theta}_i = \hat{\Omega}_i^{cl}. \tag{11.3}$$

The spin fluctuations $\delta\hat{\Omega} = \hat{\Omega} - \hat{\Omega}^{cl}$ are parametrized by two sets of variables given by the projections

$$\begin{aligned}
\mathbf{q} = \{q_i\}_1^{\mathcal{N}} &\equiv \{\delta\hat{\Omega}_i \cdot \hat{\phi}_i\}_1^{\mathcal{N}}, \\
\mathbf{p} = \{p_i\}_1^{\mathcal{N}} &\equiv \{S\delta\hat{\Omega}_i \cdot \hat{\theta}_i\}_1^{\mathcal{N}}.
\end{aligned} \tag{11.4}$$

We assume that the path integral is dominated by small fluctuations, i.e., $|\delta\hat{\Omega}| \ll 1$. Thus, we replace its measure by the phase-space measure

$$\mathcal{D}\delta\hat{\Omega} \rightarrow \nu\mathcal{D}\mathbf{q}\,\mathcal{D}\mathbf{p}, \qquad \nu = \left(\frac{2S+1}{4\pi S}\right)^{N_\epsilon \mathcal{N}}. \tag{11.5}$$

N_ϵ is the number of timesteps, and thus ν is an infinite (but uninteresting) normalization constant. The domain of integration for the p's and q's may

---

[1]This has been called in the literature *"order due to disorder"*; see bibliography.

be extended to $\int_{-\infty}^{\infty}$, and we obtain a *phase-space path integral*.[2]

$$G \propto e^{iS[\hat{\Omega}^{cl}]} \int \mathcal{D}\mathbf{q} \, \mathcal{D}\mathbf{p} \, \exp\left[\frac{iS}{2} \int_0^t dt' \, (\mathbf{q}, \mathbf{p})\mathcal{L}^{(2)} \begin{pmatrix} \mathbf{q} \\ \mathbf{p} \end{pmatrix}\right] [1 + \mathcal{O}(\mathbf{p}, \mathbf{q})^3],$$

$$(11.6)$$

where $\mathcal{L}^{(2)}$ is the spin wave Lagrangian matrix, expanded about a fixed ground state configuration $\hat{\Omega}^{cl}$. The cubic and higher-order fluctuations introduce corrections which are higher order in S^{-1}. Since $\frac{d}{dt}\hat{\Omega}^{cl} = 0$, the lowest-order contribution to the Berry phase is

$$S \sum_i \omega_i \approx S \sum_i \int_0^t dt' \left(\hat{\Omega}^{cl} \cdot \delta\hat{\Omega} \times \delta\dot{\hat{\Omega}}\right)$$

$$= \frac{1}{2} \int_0^t dt' \, (\mathbf{p} \cdot \dot{\mathbf{q}} - \dot{\mathbf{p}} \cdot \mathbf{q}),$$

$$(11.7)$$

where we have used (10.34). The kinetic term (11.7) implies that $\mathbf{q}$ and $\mathbf{p}$ play the role of coordinates and canonical momenta, respectively.

The second-order expansion of the Hamiltonian yields the spin wave Hamiltonian matrix

$$H - H[\hat{\Omega}^{cl}] \approx \frac{1}{2}(\mathbf{q}, \mathbf{p})H^{(2)} \begin{pmatrix} \mathbf{q} \\ \mathbf{p} \end{pmatrix},$$

$$H^{(2)} = \begin{pmatrix} K & P \\ P^T & M^{-1} \end{pmatrix}.$$

$$(11.8)$$

$H^{(2)}$ is a dynamical matrix of coupled harmonic oscillators, where K and M are the mass and force constant matrices, respectively,

$$K = \frac{\partial^2 H}{\partial\mathbf{q}\partial\mathbf{q}}\bigg|_{\mathbf{q}=\mathbf{p}=0},$$

$$M^{-1} = \frac{\partial^2 H}{\partial\mathbf{p}\partial\mathbf{p}}\bigg|_{\mathbf{q}=\mathbf{p}=0}.$$

$$(11.9)$$

P couples coordinates and momenta,

$$P = \frac{\partial^2 H}{\partial\mathbf{p}\partial\mathbf{q}}\bigg|_{\mathbf{q}=\mathbf{p}=0}.$$

$$(11.10)$$

Combining (11.7) and (11.8), we obtain the small oscillations contribution to the action

$$\mathcal{S}^{(2)} \approx \mathcal{S}_0 + \int_0^t dt' \, (\mathbf{q}, \mathbf{p})\mathcal{L}^{(2)} \begin{pmatrix} \mathbf{q} \\ \mathbf{p} \end{pmatrix},$$

$$(11.11)$$

---

[2]See Schulman, bibliography of Chapter 10.

where

$$L^{(2)} = -\begin{pmatrix} K & \partial_t + P \\ -\partial_t + P^T & M^{-1} \end{pmatrix}. \tag{11.12}$$

The spin wave modes $(q_{\mathbf{k},\alpha}, p_{\mathbf{k},\alpha})$ are obtained by solving the classical equations of motion for small oscillations,

$$\frac{\delta S^{(2)}(\mathbf{p}, \mathbf{q})}{\delta \hat{\Omega}} = L^{(2)} \begin{pmatrix} q_{\mathbf{k},\alpha} \\ p_{\mathbf{k},\alpha} \end{pmatrix} \exp\left(i\omega_{\mathbf{k},\alpha}\, t\right) = 0. \tag{11.13}$$

If $\hat{\Omega}^{cl}$ (and thus $L^{(2)}$) are periodic on the lattice, the spin waves are labelled by momentum $\mathbf{k}$ and band index α. The spin wave frequencies are given by the characteristic equation

$$\det\begin{pmatrix} K & i\omega + P \\ -i\omega + P^T & M^{-1} \end{pmatrix}_{\omega = \omega_{\mathbf{k},\alpha}} = 0. \tag{11.14}$$

The harmonic spin waves are noninteracting bosons. Their eigenmodes and energies allow us to calculate all thermodynamic averages and dynamical response functions. In particular, the spin wave correction to the free energy is given by

$$\begin{aligned} F^{(2)} &= \frac{T}{2} \ln \det \mathcal{L}^{(2)} \\ &= T \sum_{\mathbf{k},\alpha} \ln \sinh \left(\frac{\omega_{\mathbf{k}}}{2T}\right). \end{aligned} \tag{11.15}$$

We can transform the spin wave to Bose coherent states variables (see Appendix C) as follows:

$$\begin{pmatrix} z_i(\tau) \\ z_i(\tau)^* \end{pmatrix} = \frac{1}{\sqrt{2}} \begin{pmatrix} 1 & i \\ 1 & -i \end{pmatrix} \begin{pmatrix} q_i(\tau) \\ p_i(\tau) \end{pmatrix}. \tag{11.16}$$

The z variables transform the phase-space path integral (11.6) into a Bose coherent states path integral.[3] The measures are related by

$$\mathcal{D}\mathbf{p}\,\mathcal{D}\mathbf{q} = \mathcal{D}^2\mathbf{z} \tag{11.17}$$

and the kinetic term is

$$\frac{i}{2} \int_0^t dt' (\mathbf{p} \cdot \dot{\mathbf{q}} - \dot{\mathbf{p}} \cdot \mathbf{q}) = \frac{1}{2} \int_0^t dt' (\mathbf{z}^* \dot{\mathbf{z}} - \dot{\mathbf{z}}^* \mathbf{z}). \tag{11.18}$$

Thus, the Green's function is given by

$$G = \int \mathcal{D}^2\mathbf{z}\, \exp\left[i\frac{1}{2} \int_0^t dt' (\mathbf{z}^* \dot{\mathbf{z}} - \dot{\mathbf{z}}^* \mathbf{z}) - H[\mathbf{z}^*, \mathbf{z}]\right], \tag{11.19}$$

---

[3]See Schulman's book for the real-time coherent states path integral.

where the harmonic Bose Hamiltonian is

$$H[\mathbf{z}^*, \mathbf{z}] \approx \frac{1}{2}(\mathbf{z}^*, \mathbf{z})\tilde{H}^{(2)}\begin{pmatrix} \mathbf{z} \\ \mathbf{z}^* \end{pmatrix} ,$$

$$\tilde{H}^{(2)} = \begin{pmatrix} \frac{1}{2}(K + M^{-1}) & \frac{1}{2}(K - M^{-1}) + iP \\ \frac{1}{2}(K - M^{-1}) - iP & \frac{1}{2}(K + M^{-1}) \end{pmatrix}, \quad (11.20)$$

which has both normal (i.e., z^*z) and anomalous (z^*z^* and zz) terms.

Using the transformation from real to imaginary time,

$$t' \rightarrow -i\tau,$$
$$t \rightarrow -i\beta. \quad (11.21)$$

The path integral for G in (11.19) is transformed into the partition function (10.10). The spin wave contribution to the partition function (defined in (10.23)) is

$$Z' \approx \oint \mathcal{D}^2 z_\omega \exp\left\{-\beta \sum_{\omega_n} \left[i\omega_n z^* z - \frac{1}{2}(\mathbf{z}^*, \mathbf{z})\tilde{H}^{(2)}\begin{pmatrix} \mathbf{z} \\ \mathbf{z}^* \end{pmatrix}\right]\right\}$$

$$= \prod_n \det\left[\begin{pmatrix} i\omega_n & 0 \\ 0 & -i\omega_n \end{pmatrix} - \tilde{H}^{(2)}\right]^{-1}$$

$$\omega_n = 2n\pi/\beta. \quad (11.22)$$

ω_n are Bose Matsubara frequencies, and $\sum_\omega$ is a discrete Matsubara sum defined in Appendix D, Section D.3.

Working backwards from the coherent state Hamiltonian (11.20) to a second quantized Bose Hamiltonian, we obtain

$$\mathcal{H}^{(2)} = \frac{1}{2}(b^\dagger, b) H\begin{pmatrix} \frac{1}{2}(K + M^{-1}) & \frac{1}{2}(K - M^{-1}) + iP \\ \frac{1}{2}(K - M^{-1}) - iP & \frac{1}{2}(K + M^{-1}) \end{pmatrix}\begin{pmatrix} b \\ b^\dagger \end{pmatrix} + E'_0,$$
$$(11.23)$$

which contains both normal and anomalous terms. E'_0 is an unknown constant. The action (11.22) contains no information about the correct way to order the operators b and $b^\dagger$ in $\mathcal{H}^{(2)}$, hence the ambiguity in the constant E'_0. In the next section we shall resolve this ambiguity by using Holstein and Primakoff bosons.

Equations (11.13) or (11.23) are of general use. They determine the spin wave modes of *any* Heisenberg model once $\hat{\Omega}^{cl}$ has been specified. In the following, we restrict ourselves to the simplest models of Heisenberg ferromagnets and antiferromagnets on cubic lattices in one, two, and three dimensions:

$$H = \pm|J|S^2 \sum_{\langle ij \rangle} \hat{\Omega}_i \cdot \hat{\Omega}_j$$

$$= JS^2 \sum_{\langle ij \rangle} [\cos\theta_i \cos\theta_j + \sin\theta_i \sin\theta_j \cos(\phi_i - \phi_j)], \quad (11.24)$$

where $\langle ij \rangle$ denotes a nearest neighbor bond on the cubic lattice. The spin wave expansions of these Hamiltonians are particularly simple since their classical ground states are either uniform ferromagnets or Néel antiferromagnets.

11.1.1   THE FERROMAGNET

Assuming that a spin wave expansion of the path integral is valid about the ordered classical ground state, we can polarize the spins (with a small magnetic field, or boundary conditions) to point in the $\hat{x}$ direction, i.e.,

$$\hat{\Omega}^{cl} = (\theta^{cl}, \phi^{cl}) = (\pi/2, 0). \tag{11.25}$$

The fluctuations are parametrized by

$$\begin{aligned} q_i &= \phi_i, \\ p_i &= S\cos\theta_i. \end{aligned} \tag{11.26}$$

We expand the ferromagnetic $(-|J|)$ Hamiltonian (11.24) for small

$$|\cos\theta| << 1 , \qquad |\phi_i| << \pi \tag{11.27}$$

and obtain to second (harmonic) order

$$H \approx -\frac{Nz}{2}|J|S^2 + \frac{1}{2}|J|S^2 \sum_{\langle ij \rangle} \left[\frac{(p_i - p_j)^2}{S^2} + (q_i - q_j)^2 \right] + \dots, \tag{11.28}$$

where N is the number of sites and $z = 2d$ is the coordination number in d dimensions. By (11.9) and (11.10) we find that $P = 0$ and that

$$\begin{aligned} M_{ij}^{-1} &= -z|J|\nabla_{ij}^2 , \\ K_{ij} &= -z|J|S^2\nabla_{ij}^2, \end{aligned} \tag{11.29}$$

where the *lattice Laplacian* is given by the matrix

$$\nabla_{ij}^2 = z^{-1} \sum_{\eta} (\delta_{i+\eta,j} - \delta_{i,j}). \tag{11.30}$$

η are the nearest neighbor vectors. The eigenvalues of ∇^2 are given by

$$\begin{aligned} \nabla_{\mathbf{k}}^2 &= N^{-1} \sum_{ij} e^{-\mathbf{k}\cdot(\mathbf{x}_i - \mathbf{x}_j)} \nabla_{ij}^2 \\ &= z^{-1} \sum_{\eta} (e^{i\mathbf{k}\eta} - 1) \equiv \gamma_{\mathbf{k}} - 1 , \end{aligned} \tag{11.31}$$

which defines the "tight binding" function $\gamma_{\mathbf{k}}$. Using (11.29) with (11.14), we have

$$\det \begin{pmatrix} -z|J|S^2(\gamma_{\mathbf{k}} - 1) & i\omega_{\mathbf{k}} \\ -i\omega_{\mathbf{k}} & -z|J|(\gamma_{\mathbf{k}} - 1) \end{pmatrix} = 0, \tag{11.32}$$

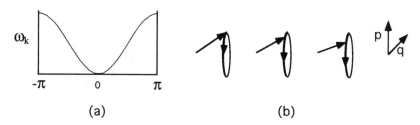

FIGURE 11.2. Ferromagnetic spin waves. (a) Dispersion in $d = 1$. (b) Polarization for $k \approx 0$.

whose zeros are at

$$\omega_{\mathbf{k}} \equiv z|J|S(1 - \gamma_{\mathbf{k}}). \tag{11.33}$$

As shown in Exercise 1 and Fig. 11.2, the spin wave is described by a time dependent precession of all spins about $\hat{x}$ with angular frequency $\omega_{\mathbf{k}}$.

11.1.2   THE ANTIFERROMAGNET

We assume the path integral can be expanded about the classical Néel state. Without loss of generality, we choose the Néel state to point in the $\hat{x}$ $(-\hat{x})$ directions in sublattice A (B):

$$(\theta_i^{cl}, \phi_i^{cl}) = \begin{cases} (\pi/2, 0) & i \in A \\ (\pi/2, -\pi) & i \in B \end{cases}. \tag{11.34}$$

(Here, it is convenient to define the singularity of vector potential to be far from both $\hat{x}$ and $-\hat{x}$, i.e., use the gauge potential (10.18), which has a singularity only at the south pole.) The small fluctuations are defined by

$$q_i = \begin{cases} \phi_i & i \in A \\ \phi_i + \pi & i \in B \end{cases},$$

$$p_i = S\cos\theta_i. \tag{11.35}$$

The second-order expansion of the antiferromagnetic $(+|J|)$ Hamiltonian (11.24) is

$$H \approx -\frac{\mathcal{N}z}{2}|J|S^2 + \frac{1}{2}|J|S^2 \sum_{\langle ij \rangle} \left[\frac{(p_i + p_j)^2}{S^2} + (q_i - q_j)^2 \right] + \ldots. \tag{11.36}$$

The mass and force constant matrices are

$$M_{ij}^{-1} = zJ\left(\nabla_{ij}^2 + 2\delta_{ij}\right),$$

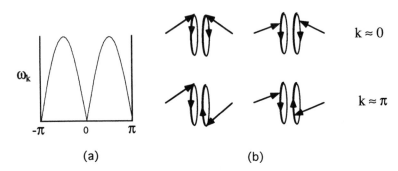

FIGURE 11.3. Antiferromagnetic spin waves. (a) Dispersion in $d = 1$. (b) Polarizations of two degenerate modes.

$$K_{ij} \quad = \quad = -zJS^2 \nabla_{ij}^2, \tag{11.37}$$

which can be simultaneously diagonalized in Fourier space as

$$M_{\mathbf{k}}^{-1} \quad = \quad zJ\left(1 + \gamma_{\mathbf{k}}\right),$$
$$K_{\mathbf{k}} \quad = \quad zJS^2\left(1 - \gamma_{\mathbf{k}}\right). \tag{11.38}$$

The spin wave spectrum is found by solving (11.14),

$$\det\begin{pmatrix} -z|J|S^2(\gamma_{\mathbf{k}} - 1) & i\omega_{\mathbf{k}} \\ -i\omega_{\mathbf{k}} & z|J|(\gamma_{\mathbf{k}} + 1) \end{pmatrix} = 0, \tag{11.39}$$

which yields for the cubic lattice the two modes

$$\omega_{\mathbf{k}} \equiv zJS\sqrt{1 - \gamma_{\mathbf{k}}^2} \sim c|\mathbf{k} - \mathbf{k}_c|, \tag{11.40}$$

where $\mathbf{k}_c = 0, \vec{\pi}$, and $\vec{\pi} = (\pi, \pi, \ldots)$. The spin wave velocity for a cubic lattice in d dimensions is

$$c = \sqrt{d}JS. \tag{11.41}$$

The antiferromagnetic spin waves are unit vectors which precess about the classical directions $\pm\hat{x}$. As worked out in Exercise 2 and depicted in Fig. 11.3, the spins precess in opposite directions on the two sublattices. Unlike ferromagnetic spin waves (11.33), there are two degenerate antiferromagnetic spin wave modes whose frequencies vanish at $\mathbf{k} \to 0$ and $\mathbf{k} \to \vec{\pi}$, respectively.

11.2   Spin Waves: Holstein–Primakoff Approach

We have seen in the previous section that the low-order spin waves can be written as noninteracting bosons. For a quantitative quantum theory, the correct order of the Bose operators in the Hamiltonian must be determined. To that end, we use Holstein and Primakoff (HP) bosons to represent the quantum model. HP bosons we previously defined in (7.1) with respect to the $\hat{z}$ direction in spin space. Here we shall generalize this definition for arbitrary classical configuration $\hat{\Omega}_i^{cl}$ which minimizes $H[\hat{\Omega}]$. Using $\hat{\Omega}^{cl}$ we can define three basis vectors $(\mathbf{e}^1, \mathbf{e}^2, \hat{\Omega}^{cl})$ at every site such that

$$\mathbf{e}_i^1 \times \mathbf{e}_i^2 = \hat{\Omega}_i^{cl}. \tag{11.42}$$

The raising and lowering operators in this coordinate frame are

$$S^{\pm} = \mathbf{S} \cdot \mathbf{e}^1 \pm i\mathbf{S} \cdot \mathbf{e}^2. \tag{11.43}$$

Similar to (7.1), the spin components (with respect to the triad in (11.42)) can be represented by HP bosons:

$$
\begin{aligned}
S^+ &= \left(\sqrt{2S - n_b}\right) b, \\
S^- &= b^\dagger \sqrt{2S - n_b}, \\
\mathbf{S} \cdot \hat{\Omega}^{cl} &= -n_b + S .
\end{aligned}
\tag{11.44}
$$

This representation is exact in the Hilbert subspace of $n_b \leq S$. However, the square-root function represents an infinite power series of number operators multiplied by factors of $1/S$. The truncatation of the series to low orders can be justified if one can show a posteriori that

$$\langle n_b \rangle \ll 2S , \tag{11.45}$$

i.e., that the spin fluctuations about the classical directions are small.

Next the spin operators in the Hamiltonian are substituted by expressions (11.42), and terms of the same order in $1/S$ are combined. Terms up to quadratic order in the bosons constitute the spin wave Hamiltonian.

11.2.1 THE FERROMAGNET

Without loss of generality, we choose the classical direction in the $\hat{z}$ direction and substitute the HP operators into the ferromagnetic Hamiltonian

$$
\begin{aligned}
\mathcal{H} &= -|J| \sum_{\langle ij \rangle} \mathbf{S}_i \cdot \mathbf{S}_j, \\
&= -S^2|J|\mathcal{N}z/2 - |J| \sum_{\langle ij \rangle} \left[Sb_i^\dagger \sqrt{1 - n_i/2S} \sqrt{1 - n_j/2S}\, b_j \right. \\
&\quad \left. -\frac{1}{2}S(n_i + n_j) + \frac{1}{2}n_i.n_j \right].
\end{aligned}
\tag{11.46}
$$

By expanding the square roots using (7.4), and collecting terms up to order $1/S^0$, we obtain

$$\mathcal{H} \approx -S^2|J|\mathcal{N}z/2 + H_1 + H_2 + \mathcal{O}(1/S),$$

$$H_1 = \sum_{\mathbf{k}} \omega_{\mathbf{k}} b_{\mathbf{k}}^\dagger b_{\mathbf{k}},$$

$$H_2 = |J|/4 \sum_{\langle ij \rangle} \left[b_i^\dagger b_j^\dagger (b_i - b_j)^2 + (b_i^\dagger - b_j^\dagger)^2 b_i b_j \right], \qquad (11.47)$$

where

$$b_{\mathbf{k}} = \frac{1}{\sqrt{\mathcal{N}}} \sum_i e^{-i\mathbf{k}\cdot\mathbf{x}_i} b_i \qquad (11.48)$$

and

$$\omega_{\mathbf{k}} = S|J|z \left(1 - z^{-1} \sum_{j,<ij>} e^{i(\mathbf{x}_j - \mathbf{x}_i)\mathbf{k}} \right) \equiv S|J|z \left(1 - \gamma_{\mathbf{k}} \right), \quad (11.49)$$

which agrees with (11.33). H_1 describes noninteracting spin waves with dispersion $\omega_{\mathbf{k}}$. H_2 and the higher-order terms describe interactions between spin waves, which can be treated by perturbation theory or by mean field approximations. At $T = 0$, it is clear from (11.47) that the energy is equal to the classical energy

$$E_0 = -S^2|J|\mathcal{N}z/2. \qquad (11.50)$$

This is in agreement with Theorem 5.1 in Chapter 5, which proves that the ground state of the quantum ferromagnet is the classical ground state.

The long-wavelength limit of the ferromagnetic spin wave dispersion vanishes as

$$\omega_{\mathbf{k}} \sim S|J||\mathbf{k}|^2 . \qquad (11.51)$$

This is the gapless *Goldstone mode*, which is a consequence of the broken symmetry of the ferromagnetic ground state, as shown in Section 9.2. The Goldstone mode dominates the low-temperature and long-wavelength correlations of the ferromagnet. The lowest-order corrections to the ground state magnetization at finite temperatures is given by

$$\Delta m_0 = \frac{1}{\mathcal{N}} \langle S_{tot}^z \rangle - S$$

$$= -\langle n_i \rangle = -\frac{1}{\mathcal{N}} \sum_{\mathbf{k}} n_{\mathbf{k}}, \qquad (11.52)$$

where the Bose–Einstein occupation number is (see (A.25))

$$n_{\mathbf{k}} = \frac{1}{e^{\omega_{\mathbf{k}}/T} - 1} . \qquad (11.53)$$

The asymptotic low-T behavior of the sum in (11.52) is found by introducing a small "*infrared*" cutoff k_0. We choose an additional small but finite momentum $\bar{k} > k_0$ which is well inside the region where (11.51) holds, i.e.,

$$\omega_{\bar{k}} << T << |J|S . \tag{11.54}$$

Now we break up the sum in (11.52) into $k_0 < |\mathbf{k}| < \bar{k}$ and $|\mathbf{k}| \geq \bar{k}$ and find that

$$\Delta m_0 \approx -\int_{k_0}^{\bar{k}} \frac{dk\, k^{d-1}}{(2\pi)^d} \frac{T}{JSk^2} - \mathcal{N}^{-1} \sum_{|\mathbf{k}|>\bar{k}} \frac{1}{\exp[\omega_{\mathbf{k}}/T] - 1}. \tag{11.55}$$

For $T > 0$ in $d = 1, 2$, the first integral diverges at low k_0 as

$$\Delta m_0 = \propto \begin{cases} -\frac{t}{k_0} + \dots & d = 1 \\ t \log k_0 + \dots & d = 2 \end{cases}, \tag{11.56}$$

where $t = T/(JS)$. The second sum in (11.55) is finite, and does not affect the infrared singularity. The most important conclusion emerging from (11.56) is that the magnetization correction due to fluctuations diverges as the infrared cutoff k_0 vanishes in one and two dimensions. Therefore, our initial assumption that the spin fluctuations $\langle (S_i^z - S)^2 \rangle$ are small is found to be *wrong* for these cases. Thus, the truncation of the expansion of the HP operators at low orders is unjustified. The breakdown of spin wave theory is consistent with Mermin and Wagner's Theorem 6.2, which does rule out a finite spontaneous magnetization in $d=1,2$ at all nonzero temperatures.

In $d = 3$ there are no infrared divergences. The leading temperature dependence of M can be calculated by writing the Bose function (11.53) as a geometric sum and integrating over momenta up to infinity as follows:

$$\begin{aligned} \Delta m_0^{d=3} &= -\int_{k_0}^{\bar{k}} \frac{dk\, k^2}{2\pi^2} \sum_{n=1}^{\infty} \exp[-nk^2/t] \; [1 + \mathcal{O}(t)] \\ &\approx -\frac{1}{8} \left(\frac{t}{\pi}\right)^{3/2} \sum_{n=1}^{\infty} n^{-3/2} = -\frac{1}{8} \left(\frac{t}{\pi}\right)^{3/2} \zeta(3/2), \end{aligned} \tag{11.57}$$

where ζ is the Riemann zeta function $\zeta(s) = \sum_n n^{-s}$. Thus, in three dimensions the leading temperature correction to the ordered moment is proportional to $-T^{3/2}$.

11.2.2 THE ANTIFERROMAGNET

Here we choose the classical Néel state to be in the $\hat{z}$ and $-\hat{z}$ directions on sublattices A and B, respectively. We define the rotated spins as $\tilde{\mathbf{S}}$:

$$j \in B ,$$

$$\begin{aligned}
\tilde{S}_j^z &= -S_j^z, \\
\tilde{S}_j^x &= S_j^x, \\
\tilde{S}_j^y &= -S_j^y.
\end{aligned} \tag{11.58}$$

It is clear that $\tilde{S}^\alpha$ obey the same commutation relations as S^α and therefore can be represented by Holstein–Primakoff bosons. Using the sublattice rotated representation, the antiferromagnetic Hamiltonian reads

$$\begin{aligned}
\mathcal{H} &= -|J| \sum_{\langle ij \rangle} S_i^z \cdot \tilde{S}_j^z \\
&\quad + \frac{1}{2} |J| \sum_{\langle ij \rangle} \left(S_i^+ \tilde{S}_j^+ + S_i^- \tilde{S}_j^- \right).
\end{aligned} \tag{11.59}$$

Using the HP boson representation (11.44) for $\mathbf{S}$ on sublattice A and $\tilde{\mathbf{S}}$ for sublattice B, one obtains

$$\begin{aligned}
\mathcal{H} &= -S^2 J \mathcal{N} z / 2 + \mathcal{H}_1 + \mathcal{H}_2 + \mathcal{O}(1/S), \\
\mathcal{H}_1 &= JSz \sum_{\mathbf{k}} \left[b_{\mathbf{k}}^\dagger b_{\mathbf{k}} + \frac{\gamma_{\mathbf{k}}}{2} (b_{\mathbf{k}}^\dagger b_{-\mathbf{k}}^\dagger + b_{\mathbf{k}} b_{-\mathbf{k}}) \right], \\
\mathcal{H}_2 &= -\frac{J}{4} \sum_{\langle ij \rangle} \left[b_i^\dagger (b_j^\dagger + b_i)^2 b_j + b_j (b_i^\dagger + b_j)^2 b_i \right].
\end{aligned} \tag{11.60}$$

$\mathcal{H}_1$ is a quadratic Hamiltonian which includes normal and anomalous terms (e.g., such as $a^\dagger a^\dagger$). We can diagonalize $\mathcal{H}_1$ by a *Boguliubov* transformation. Let us define the spin wave operator $\alpha_{\mathbf{k}}$ such that

$$\begin{aligned}
\alpha_{\mathbf{k}} &= \cosh \theta_{\mathbf{k}} a_{\mathbf{k}} - \sinh \theta_{\mathbf{k}} a_{-\mathbf{k}}^\dagger, \\
a_{\mathbf{k}} &= \cosh \theta_{\mathbf{k}} \alpha_{\mathbf{k}} + \sinh \theta_{\mathbf{k}} \alpha_{-\mathbf{k}}^\dagger.
\end{aligned} \tag{11.61}$$

The parameters $\theta_{\mathbf{k}}$ are real, and even in $\mathbf{k} \to -\mathbf{k}$. Equation (11.61) is a canonical transformation since

$$\begin{aligned}
\left[\alpha_{\mathbf{k}}, \alpha_{\mathbf{k}'}^\dagger \right] &= \delta_{\mathbf{k}\mathbf{k}'}, \\
\left[\alpha_{\mathbf{k}}, \alpha_{\mathbf{k}'} \right] &= \left[\alpha_{\mathbf{k}}^\dagger, \alpha_{\mathbf{k}'}^\dagger \right] = 0.
\end{aligned} \tag{11.62}$$

In terms of spin wave operators, $\mathcal{H}_1$ is given by

$$\begin{aligned}
\mathcal{H}_1 &= |J| Sz \sum_{\mathbf{k}} \left[(\cosh 2\theta_{\mathbf{k}} + \gamma_{\mathbf{k}} \sinh 2\theta_{\mathbf{k}}) \alpha_{\mathbf{k}}^\dagger \alpha_{\mathbf{k}} \right. \\
&\quad + \frac{1}{2} (\sinh 2\theta_{\mathbf{k}} + \gamma_{\mathbf{k}} \cosh 2\theta_{\mathbf{k}}) (\alpha_{\mathbf{k}}^\dagger \alpha_{-\mathbf{k}}^\dagger + \alpha_{\mathbf{k}} \alpha_{-\mathbf{k}}) \\
&\quad \left. + \sinh^2 \theta_{\mathbf{k}} + \frac{\gamma_{\mathbf{k}}}{2} \sinh 2\theta_{\mathbf{k}} \right].
\end{aligned} \tag{11.63}$$

Now, we choose $\theta_{\mathbf{k}}$ so that the anomalous terms $\alpha^\dagger \alpha^\dagger$ and $\alpha \alpha$ vanish. This amounts to the condition

$$\tanh 2\theta_{\mathbf{k}} = -\gamma_{\mathbf{k}}, \tag{11.64}$$

which yields

$$\mathcal{H}_1 = \sum_{\mathbf{k}} \omega_{\mathbf{k}} \, (\alpha^\dagger_{\mathbf{k}} \alpha_{\mathbf{k}} + \tfrac{1}{2}) \; - \; \frac{JSz\mathcal{N}}{2},$$

$$\omega_{\mathbf{k}} = |J|Sz\sqrt{1 - \gamma_{\mathbf{k}}^2}, \tag{11.65}$$

which agrees with (11.40). By (11.65) we see that unlike the case of the ferromagnet (11.50), $\mathcal{H}_1$ contains a quantum zero-point energy of size

$$E_0' = E_0 - \left(-\frac{z}{2}\mathcal{N}|J|S^2\right) = \frac{1}{2}\sum_{\mathbf{k}} |J|Sz\left(\sqrt{1 - \gamma_{\mathbf{k}}^2} - 1\right). \tag{11.66}$$

Note that E_0' is negative, i.e., quantum fluctuations *reduce* the energy of the antiferromagnet. We can understand this phenomenon by considering the simplest case of two sites. While the classical ground state energy is $-JS^2$, the quantum energy is considerably lower: $-JS(S+1)$.

The ground state of the quantum antiferromagnet is the vaccuum of the α bosons. The Boguliubov transformation introduces arbitrary numbers of HP bosons in the ground state. In the spin language, the ground state admixes configurations with arbitrary numbers of spin-flips relative to the Néel state. This reduces the long-range Néel order at $T = 0$.

Near $\mathbf{k} = 0$, and near $\mathbf{k} = \vec{\pi} = (\pi, \pi, ...)$, the spin wave spectrum vanishes as

$$\omega_{\mathbf{k}} \sim \begin{cases} JS\sqrt{2z} \, |\mathbf{k}| & |\mathbf{k}| \approx 0 \\ JS\sqrt{2z} \, |\mathbf{k} - (\pi, \pi, ...)| & \mathbf{k} \approx (\pi, \pi, ...) \end{cases}. \tag{11.67}$$

The order parameter is the staggered magnetization. Its low-order corrections are given by $\mathcal{H}_1$ as follows:

$$\begin{aligned} \Delta m_0^s &= \frac{1}{\mathcal{N}}\langle \sum_i e^{i\vec{\pi}\mathbf{x}_i} S_i^z \rangle - S \\ &= -\frac{1}{\mathcal{N}}\langle \sum_i a_i^\dagger a_i \rangle \\ &= +\frac{1}{2} - \frac{1}{\mathcal{N}}\sum_{\mathbf{k}}(n_{\mathbf{k}} + \tfrac{1}{2})\frac{1}{\sqrt{1 - \gamma_{\mathbf{k}}^2}}. \end{aligned} \tag{11.68}$$

As in the case of the ferromagnet, the truncation of the HP expansion to quadratic order is justified only when $\Delta m_0^s \ll S$.

The leading infrared singularities in Δm_0^s at low temperatures is evaluated by expanding the Bose function near $\omega_{\mathbf{k}} \approx 0$. Using k_0 to cut off the momentum sum near $\mathbf{k} \approx 0$ and $\mathbf{k} \approx \vec{\pi}$, we obtain

$$\Delta m_0 \sim \begin{cases} \frac{c_1}{k_0} & d = 1 \\ -c_2 + t \log k_0 & d = 2 \,, \\ -c_3 + c_3' t^2 & d = 3 \end{cases} \tag{11.69}$$

where

$$\begin{aligned} c_d &= \frac{1}{2N} \sum_{\mathbf{k}} \frac{1}{\sqrt{1 - \gamma_{\mathbf{k}}^2}} - 1, \quad d = 2, 3, \\ c_3' &= 6^{-5/2}/2, \\ t &= \frac{T}{|J| S \sqrt{d}}. \end{aligned} \tag{11.70}$$

As in the case of the ferromagnet, the truncation of the HP expansion is justified only when we have true long-range order and $\Delta m_0^s \ll S$. We note that the staggered magnetization correction actually diverges as $k_0 \to 0$ in one dimension. This signals the failure of spin wave theory and no long-range order in the ground state. Later, we shall see that the absence of long-range order in the ground state of the d-dimensional quantum model is related to the Mermin and Wagner theorem for the *classical* Heisenberg model in $d + 1$ dimensions at finite temperatures.

We also note that by (11.69), the existence of long-range order in the ground state for $d = 2, 3$ depends on the relative sizes of c' and S. $d=3$ is the lowest dimension where long range spin order can possibly exist at finite temperatures.

11.3 Exercises

1. Use (11.13) for the ferromagnetic spin wave modes, and show that the directions of the precessing spins at $k \approx 0$ are given by Fig. 11.2.

2. Repeat the previous Exercise for the two spin wave modes at $k = 0$ and $k = \pi$ of the one-dimensional antiferromagnet, and compare them to Fig. 11.3.

3. The Néel vector for a nearest neighbor bond is defined as

$$\hat{\mathbf{n}} = \frac{1}{2}(\hat{\Omega}_i - \hat{\Omega}_j). \tag{11.71}$$

Show that $\hat{\mathbf{n}}$ approximately executes periodic planar oscillations. Find the planes of polarizations of $\hat{\mathbf{n}}$ for $\mathbf{k} = 0$ and $\mathbf{k} = \vec{\pi}$.

4. Consider the *ferrimagnet* on a cubic lattice in which the spins on sublattice A, of size S_A, couple antiferromagnetically to their neighbors on sublattice B, which have size S_B. Following subsection 11.1.2, use spin wave theory to compute the dispersions of the elementary excitations.

5. The ground state of the antiferromagnetic spin wave Hamiltonian $\mathcal{H}_1$ (see (11.60)) must obey

$$\alpha_{\mathbf{k}} \, \Psi_0 \; = \; 0 \, , \qquad \forall \, \mathbf{k} . \tag{11.72}$$

Using the definition of $\alpha_{\mathbf{k}}$ in (11.61), prove that

$$\Psi_0 \; = \; \nu \, \exp \left[\frac{1}{2} \sum_{\mathbf{k}} u_{\mathbf{k}} b^{\dagger}_{\mathbf{k}} b^{\dagger}_{-\mathbf{k}} \right] \, |0\rangle \, ,$$

$$u_{\mathbf{k}} \; = \; \tanh \theta_{\mathbf{k}} \, . \tag{11.73}$$

Find the normalization constant ν. *Note:* Ψ_0 *in (11.73) is not a pure spin state since it has contributions from nonphysical states with $n_b > 2S$. Proper variational spin states can be defined by replacing the HP bosons by spin raising and lowering operators.*

6. Evaluate the asymptotic decay of u_{ij} for large $|\mathbf{x}_i - \mathbf{x}_j|$,

$$u_{ij} \; = \; \frac{1}{\mathcal{N}} \sum_{\mathbf{k}} u_{\mathbf{k}} \exp \left[-i \mathbf{k} \cdot (\mathbf{x}_i - \mathbf{x}_j) \right] \, , \tag{11.74}$$

where $u_{\mathbf{k}}$ is defined in (11.73). *Hint: use dimensional analysis to scale powers of $|\mathbf{x}_i - \mathbf{x}_j|$ out of the integral.*

7. Expand the isotropic spin correlation function

$$S(\mathbf{q}) \; = \; \mathcal{N}^{-1} \langle \mathbf{S}_{\mathbf{q}} \cdot \mathbf{S}_{-\mathbf{q}} \rangle \tag{11.75}$$

up to the four bosons terms using the HP representation for the ferromagnet. Use pairwise contractions to write $S(\mathbf{q})$ at finite temperatures as a single momentum sum.

Bibliography

Spin wave theory was originally used by

- P.W. Anderson, Phys. Rev. **83**, 1260 (1951);
- R. Kubo, Phys. Rev. **87**, 568 (1952);
- T. Oguchi, Phys. Rev. **117**, 117 (1960).

For extensive reviews on spin wave theory and applications, see

- D.C. Mattis, *Theory of Magnetism I* (Springer-Verlag, 1988);
- R.M. White, *Quantum Theory Of Magnetism* (Springer-Verlag, 1987).

Spin wave corrections to frustrated Heisenberg antiferromagnets and "Order Due to Disorder" effects are discussed by

- E.F. Shender and P.C.W. Holdsworth, *Fluctuations and Order: The New Synthesis*, edited by M.M. Millonas (MIT Press, 1994);
- C.L. Henley, Phys. Rev. Lett. **62**, 2056 (1989).

12

The Continuum
Approximation

In Chapter 11, we derived spin wave theory in a semiclassical expansion of the spin path integral. Spin wave theory, however, is not applicable to the disordered (or "rotationally symmetric") phases of the Heisenberg model since it assumes that the $O(3)$ symmetry is spontaneously broken. According to Mermin and Wagner's Theorem 6.2, there can be no spontaneously broken symmetry at $T > 0$ in one and two dimensions. Moreover, we know that for the Heisenberg antiferromagnet in one dimension, we do not expect long-range spin order even at $T = 0$.

The first question that comes to mind is: can one use the semiclassical approximation in the absence of spontaneously broken symmetry? In this chapter, we shall see that the answer is yes. A short-range classical Hamiltonian is sensitive mostly to short-range correlations. Thus, for the Heisenberg antiferromagnet, the important configurations in the semiclassical (large S) limit (i.e., "*semiclassical configurations*") have at least short-range antiferromagnetic order. At longer length scales the semiclassical configurations can deviate largely from the Néel state. Thus, we should not break the rotational symmetry of the path integral a priori, as we have done in the spin wave expansion. Instead, we shall eliminate the short length scale fluctuations and keep the full rotational symmetry of the long-wavelength semiclassical modes.

We consider the imaginary time path integral (10.10):

$$Z[j] = \int \mathcal{D}\hat{\Omega}(\tau) \, \exp\left(-\tilde{\mathcal{S}}[\hat{\Omega}]\right) \,,$$

$$\tilde{\mathcal{S}}[\hat{\Omega}] = -iS \sum_i \omega[\hat{\Omega}_i] + \int_0^\beta d\tau \, H(\tau) \,, \tag{12.1}$$

and specialize in the Heisenberg antiferromagnet

$$H[\hat{\Omega}] = \frac{1}{2}S^2 \sum_{ij} J_{ij}\hat{\Omega}_i \cdot \hat{\Omega}_j \,. \tag{12.2}$$

We shall consider here cubic lattices of dimension d with lattice constant a, number of sites $\mathcal{N}$, and even number of sites in each dimension. We use units where $\hbar = K_B = 1$. The interactions J_{ij} have the full lattice symmetry, and give rise to a Néel ground state for the classical Hamiltonian $H[\hat{\Omega}]$. We also assume that J_{ij} are short range such that

$$\frac{1}{2d} \sum_j |J_{ij}||\mathbf{x}_j - \mathbf{x}_i|^2 < \infty. \tag{12.3}$$

12.1 Haldane's Mapping

Haldane has mapped the effective long-wavelength action of the quantum Heisenberg antiferromagnet in d dimensions into the nonlinear sigma model (NLSM) in $d+1$ dimensions. The NLSM is a field theory that has been extensively studied in statistical mechanics and particle physics, and we shall discuss it in following chapters.

The essence of Haldane's mapping is the separation between short and long length scale fluctuations. This separation is made possible by a suitable choice of coordinates. The first step is to define two continuous vector fields $\hat{\mathbf{n}}$ and $\mathbf{L}$, which parametrize the spins as follows:

$$\hat{\Omega}_i = \eta_i \hat{\mathbf{n}}(\mathbf{x}_i) \sqrt{1 - \left|\frac{\mathbf{L}(\mathbf{x}_i)}{S}\right|^2} + \frac{\mathbf{L}(\mathbf{x}_i)}{S}, \tag{12.4}$$

where $\eta_i = e^{i\mathbf{X}_i \cdot \vec{\pi}}$ has opposite signs on the two sublattices, and $\hat{\mathbf{n}}$ is the unimodular *Néel field*,

$$|\hat{\mathbf{n}}(\mathbf{x}_i)| = 1. \tag{12.5}$$

$\mathbf{L}$ is the transverse *canting field*, which is chosen to obey

$$\mathbf{L}(\mathbf{x}_i) \cdot \hat{\mathbf{n}}(\mathbf{x}_i) = 0 . \tag{12.6}$$

In (12.4) it seems that we have replaced two independent degrees of freedom per site (θ_i, ϕ_i), with four variables (six degrees of freedom for $(\hat{\mathbf{n}}_i, \mathbf{L}_i)$ minus two constraints (12.5) and (12.6)). This discrepancy is resolved by fixing the number of Fourier components in the measure:

$$\mathcal{D}\hat{\Omega} = \prod_{|\mathbf{q}| \leq \Lambda_{BZ}} d\hat{\mathbf{n}}_{\mathbf{q}}\, d\mathbf{L}_{\mathbf{q}}\, \delta(\mathbf{L} \cdot \hat{\mathbf{n}}) \mathcal{J}[\hat{\mathbf{n}}, \mathbf{L}], \tag{12.7}$$

where $\mathcal{J}$ is the Jacobian of transformation (12.4), and the Fourier transform is defined as

$$\mathbf{X}_{\mathbf{q}} = \sum_i e^{-i\mathbf{q} \cdot \mathbf{x}_i} \mathbf{X}(\mathbf{x}_i) , \qquad \mathbf{X} = \hat{\mathbf{n}}, \mathbf{L} \dots, \tag{12.8}$$

Λ_{BZ} is the spherical Brillouin zone radius chosen such that

$$2\mathcal{N} = 4 \sum_{|\mathbf{q}| \leq \Lambda_{BZ}} . \tag{12.9}$$

The left- and right-hand sides of (12.9) count the degrees of freedom for the left- and right-hand sides of (12.7), respectively.

In the short-range ordered phases, Z is dominated by configurations with sizeable antiferromagnetic correlations for distances below the correlation length ξ. For large ξ/a we assume that we can find an intermediate momentum cutoff scale Λ that is much smaller than the microscopic cutoff momenta but also much larger than the inverse correlation length,

$$\xi^{-1} << \Lambda << \Lambda_{BZ}, 2\pi/R_J, \qquad (12.10)$$

where R_J is the characteristic range of $J_{i,j}$. This is equivalent to assuming that the dominant configurations in the path integral are *slowly varying* on the scale of Λ^{-1}, or that they have negligble Fourier components for

$$\left|\hat{\mathbf{n}}_{|\mathbf{q}|>\Lambda}\right| << 1. \qquad (12.11)$$

Therefore we are permitted to replace the cubic Brillouin zone of the lattice by the spherical zone in (12.7) and (12.9). Equation (12.11) is also consistent with assuming that the canting field is small,

$$|\mathbf{L}_i/S| << 1. \qquad (12.12)$$

To leading order in $|\mathbf{L}|/S$, the Jacobian of (12.7) is constant:

$$\mathcal{J} \approx S^{-\mathcal{N}}. \qquad (12.13)$$

Notice that we do not need to assume that $\mathbf{L}$ varies slowly in space. Our main assumption is encapsulated in the *existence* of a wave vector scale Λ. We must check the consistency of this assumption by evaluating $\xi(\Lambda)$ and verifying that inequality (12.10) is satisfied.

12.2   The Continuum Hamiltonian

We proceed to write the Hamiltonian using $\hat{\mathbf{n}}$ and $\mathbf{L}$. By expanding the interactions to quadratic order in $|\mathbf{L}/S|^2$, we obtain

$$\begin{aligned}
\hat{\Omega}_i \cdot \hat{\Omega}_j &\approx \eta_i \eta_j - \frac{1}{2}\eta_i \eta_j (\hat{\mathbf{n}}_i - \hat{\mathbf{n}}_j)^2 \\
&+ \left(\frac{1}{S}\right)^2 \left[\mathbf{L}_i \mathbf{L}_j - \frac{1}{2}\eta_i \eta_j (\mathbf{L}_i^2 + \mathbf{L}_j^2)\right] \\
&+ \frac{1}{S}\left(\eta_j \mathbf{L}_i \hat{\mathbf{n}}_j + \eta_i \mathbf{L}_j \hat{\mathbf{n}}_i\right) + \mathcal{O}\left(|\mathbf{L}|^2 |\hat{\mathbf{n}}_i - \hat{\mathbf{n}}_j|\right). (12.14)
\end{aligned}$$

The differences of Néel fields can be approximated by derivatives

$$\hat{\mathbf{n}}_i - \hat{\mathbf{n}}_j \approx \partial_l \hat{\mathbf{n}}(\mathbf{x}_i) \, x_{ij}^l + \frac{1}{2}(\partial_l \partial_k \hat{\mathbf{n}}) \, x_{ij}^l x_{ij}^k + \dots, \qquad (12.15)$$

where $\mathbf{x}_{ij} = \mathbf{x}_i - \mathbf{x}_j$ and summation over repeated spatial indices $l, k = 1, \ldots, d$ is assumed. Using (12.10), we can see that the higher derivatives in (12.15) are higher order in the small parameter $\Lambda R_J \ll 1$.

The first derivative cancels in the cross terms by the symmetry of the Hamiltonian, which leads to

$$\sum_{ij} J_{ij}\eta_i \mathbf{L}_j \hat{\mathbf{n}}_i \approx \sum_{ij} \eta_i J_{ij} \mathbf{L}_j \left(\hat{\mathbf{n}} + \partial_l \hat{\mathbf{n}} x_{ij}^l + \frac{1}{2}\partial_l\partial_k \hat{\mathbf{n}} x_{ij}^l x_{ij}^k \cdots \right)$$

$$\approx \frac{1}{2}\sum_{ij} J_{ij}\eta_i \mathbf{L}_j (\partial_l\partial_k \hat{\mathbf{n}}) \, x_{ij}^l x_{ij}^k \sim \mathcal{O}(\Lambda R_J)^2 \; , \quad (12.16)$$

which are negligible.

We now replace the lattice sums by integrals:

$$\sum_i F_i \rightarrow a^{-d} \int d^d x \sum_i \delta(\mathbf{x} - \mathbf{x}_i) F(\mathbf{x}), \quad (12.17)$$

and arrive at the continuum representation

$$H \approx E_0^{cl} + \frac{1}{2}\int d^d x \left[\rho_s \sum_l |\partial_l \hat{\mathbf{n}}|^2 + \int d^d x' \, (\mathbf{L}_x \chi_{xx'}^{-1} \mathbf{L}_{x'}) \right] . \quad (12.18)$$

The classical energy is

$$E_0^{cl} = S^2 \frac{1}{2}\sum_{ij} J_{ij}\eta_i\eta_j \quad (12.19)$$

and the *"stiffness constant"*[1] is

$$\rho_s = -\frac{S^2}{2d\mathcal{N}a^d}\sum_{ij} J_{ij}\eta_i\eta_j |\mathbf{x}_i - \mathbf{x}_j|^2. \quad (12.20)$$

The inverse uniform susceptibility is

$$\chi_{\mathbf{x},\mathbf{x}'}^{-1} = \frac{1}{\mathcal{N}a^d}\sum_{ij} J_{ij} \left[\delta(\mathbf{x} - \mathbf{x}_i)\delta(\mathbf{x}' - \mathbf{x}_j) \, - \delta(\mathbf{x}' - \mathbf{x})\delta(\mathbf{x} - \mathbf{x}_i)\eta_i\eta_j \right]. \quad (12.21)$$

In Fourier space, the canting field term is given by

$$\int d^d x \int d^d x' \, \mathbf{L}_{\mathbf{x}} \chi_{xx'}^{-1} \mathbf{L}_{\mathbf{x}'} = \int^{q \leq \Lambda_{BZ}} \frac{d^d q}{(2\pi)^d} \, \mathbf{L}_{\mathbf{q}} \chi^{-1}(\mathbf{q}) \mathbf{L}_{-\mathbf{q}}, \quad (12.22)$$

---

[1]The notation ρ_s originates from the continuum theory of Bose superfluids, where the stiffness constant equals the superfluid density.

where

$$\chi(\mathbf{q}) = \frac{1}{J(\mathbf{q}) - J(\vec{\pi})},$$

$$J(\mathbf{q}) = \sum_j e^{i\mathbf{q}\cdot\mathbf{x}_{ij}} J_{ij}. \tag{12.23}$$

12.3 The Kinetic Term

The kinetic Berry phase term in the spin path integral is (see (10.15))

$$-iS \sum_i \omega_i = -iS \int_0^\beta d\tau \sum_i \mathbf{A}(\hat{\Omega}_i)\dot{\hat{\Omega}}_i. \tag{12.24}$$

For convenience, we use a vector potential that is symmetric under inversion,

$$\mathbf{A}(\hat{\Omega}) = \mathbf{A}(-\hat{\Omega}), \tag{12.25}$$

such as (see (10.17))

$$\mathbf{A} = -\frac{\cos\theta}{\sin\theta}\hat{\phi}. \tag{12.26}$$

We expand (12.24) in terms of $\hat{n}$ and $\mathbf{L}$ using (10.34) and obtain

$$\begin{aligned}
-iS\sum_i \omega_i &= -iS\sum_i \eta_i\omega[\hat{\mathbf{n}}_i + \eta_i(\mathbf{L}_i/S)] \\
&= -iS\sum_i \left[\eta_i\omega[\hat{\mathbf{n}}_i] + \frac{\delta\omega}{\delta\hat{\mathbf{n}}_i}\cdot(\mathbf{L}_i/S)\right] \\
&= -i\Upsilon - i\int_0^\beta d\tau \sum_i (\hat{\mathbf{n}}_i \times \partial_\tau\hat{\mathbf{n}}_i \cdot \mathbf{L}_i), \tag{12.27}
\end{aligned}$$

where

$$\Upsilon[\hat{\mathbf{n}}] = S\sum_i \eta_i\omega[\hat{\mathbf{n}}(\mathbf{x}_i)]. \tag{12.28}$$

Υ is a *topological Berry phase* associated with the Néel field $\hat{\mathbf{n}}$. In the rotationally symmetric phases, interference between Berry phases can give rise to dramatic consequences on ground state degeneracy and low-energy spectrum. We shall discuss them in Chapter 15.

12.4 Partition Function and Correlations

The second term in (12.27) couples $\mathbf{L}$ to $\hat{\mathbf{n}} \times \partial_\tau\hat{\mathbf{n}}$. This implies that the two fields are canonical conjugates.[2] The constraint $\delta(\mathbf{L}\cdot\hat{\mathbf{n}})$ is automatically

---

[2]It is similar to the kinetic term $(\mathbf{p}\cdot\dot{\mathbf{q}} - \mathbf{q}\cdot\dot{\mathbf{p}})/2$ in (11.7).

satisfied in the second term of (12.27). By completing the square, performing the Gaussian integration on $\mathbf{L}^\perp$, and ignoring overall normalization constants, we obtain

$$
Z = \int \mathcal{D}\hat{\mathbf{n}}\, e^{i\Upsilon[\hat{\mathbf{n}}]} \exp\left(-\int_0^\beta d\tau \int_\Lambda d^d x \, \frac{\rho_s}{2} \sum_{l=1}^d |\partial_l \hat{\mathbf{n}}|^2 \right) \zeta[\hat{\mathbf{n}}] ,
$$

$$
\zeta[\hat{\mathbf{n}}] = \int_{\Lambda_{BZ}} \mathcal{D}\mathbf{L}^\perp \exp\left[-\int_0^\beta d\tau \int^\Lambda \frac{d^d q}{(2\pi)^d} (\hat{\mathbf{n}} \times \partial_\tau \hat{\mathbf{n}})_{-\mathbf{q}} \cdot \mathbf{L}_{\mathbf{q}} \right.
$$

$$
\left. -\frac{1}{2} \int_{\Lambda_{BZ}} \frac{d^d q}{(2\pi)^d} \chi^{-1}(\mathbf{q}) \mathbf{L}_{\mathbf{q}} \mathbf{L}_{-\mathbf{q}} \right]
$$

$$
\propto \exp\left[-\frac{1}{2} \int d\tau \int^\Lambda \frac{d^d q}{(2\pi)^d} \chi(\mathbf{q})(\hat{\mathbf{n}} \times \partial_\tau \hat{\mathbf{n}})_{\mathbf{q}} \cdot (\hat{\mathbf{n}} \times \partial_\tau \hat{\mathbf{n}})_{-\mathbf{q}} \right].
$$
(12.29)

By (12.23) the variation of $\chi(\mathbf{q})$ is small on the scale of Λ. Thus, we can replace it by its zero momentum limit and write

$$
\zeta[\hat{\mathbf{n}}] \approx \exp\left(-\frac{1}{2} \int_0^\beta d\tau \int_\Lambda d^d x \, \chi_0 \, |\partial_\tau \hat{\mathbf{n}}|^2 \right) ,
$$
(12.30)

where $\chi_0 = a^{-d}\chi(0)$, and $\int_\Lambda d^d x$ is understood to contain only the Fourier components of the integrand which obey $|\mathbf{q}| \leq \Lambda$. In the last line of (12.30), we used the identity

$$
|\hat{\mathbf{n}}(\mathbf{x}) \times \partial_\tau \hat{\mathbf{n}}(\mathbf{x})|^2 = |\partial_\tau \hat{\mathbf{n}}(\mathbf{x})|^2.
$$
(12.31)

Thus, we arrive at a local interaction for the $\hat{\mathbf{n}}$ fields.

By combining (12.27) and (12.30), we derive the semiclassical partition function

$$
Z \propto \int_\Lambda \mathcal{D}\hat{\mathbf{n}}\, e^{i\Upsilon[\hat{\mathbf{n}}]}
$$

$$
\times \exp\left[-\frac{1}{2} \int_0^\beta d\tau \int_\Lambda d^d x \left(\chi_0 |\partial_\tau \hat{\mathbf{n}}|^2 + \rho_s \sum_{l=1}^d |\partial_l \hat{\mathbf{n}}|^2 \right) \right].
$$
(12.32)

The *spin wave velocity* is defined as

$$
c \equiv \sqrt{\rho_s/\chi_0} .
$$
(12.33)

The Euclidean "relativistic" notation unifies the spatial and imaginary time coordinates

$$
(x_1, \ldots, x_d, c\tau) \rightarrow (x_1, \ldots, x_{d+1}).
$$
(12.34)

Using this notation, the action of Z is the *Nonlinear Sigma Model* (NLSM) of $d+1$ dimensions, with Berry phases:

$$Z \propto \int_\Lambda \mathcal{D}\hat{\mathbf{n}} \, e^{i\Upsilon} \, \exp\left(-\int d^{d+1}x \, \mathcal{L}_{NLSM}^{d+1}\right),$$

$$\mathcal{L}_{NLSM}^D = \frac{\Lambda^{D-2}}{2f_D} \sum_{\mu=1}^D \partial_\mu \hat{\mathbf{n}} \cdot \partial_\mu \hat{\mathbf{n}}, \tag{12.35}$$

where f is the dimensionless coupling constant

$$f_D = \frac{c}{\rho_s} \Lambda^{D-2}. \tag{12.36}$$

The spin correlation function of the Heisenberg antiferromagnet at long distances and at low temperatures is given by

$$\langle S_i^\alpha S_j^\alpha(\tau)\rangle_d \approx \frac{e^{i\vec{\pi}\cdot(\mathbf{X}_i - \mathbf{X}_j)}S^2}{NZ} \int_\Lambda \mathcal{D}\hat{\mathbf{n}}$$
$$\times n^\alpha(0,0) n^\alpha(\mathbf{x}_j, c\tau) e^{i\Upsilon} \exp\left(-\int_0^{c\beta} dx_{d+1} \int d^d x \, \mathcal{L}_{NLSM}^{d+1}\right). \tag{12.37}$$

The imaginary-time ground state correlations of the QHA are given by

$$Z^{-1} \sum_m |\langle \Psi_0|\hat{\mathbf{n}}(0)|\Psi_m\rangle|^2 \, \exp\left[-(E_m - E_0)\tau\right] = \langle \hat{\mathbf{n}}(0,0)\hat{\mathbf{n}}(0,\tau)\rangle. \tag{12.38}$$

The NLSM is rotationally symmetric in space-time. If the correlations decay exponentially at large distance with correlation length ξ, the imaginary time decay period is given by ξ/c. By (12.38), the lowest excitation energy Δ determines the fastest decay rate, and therefore

$$\Delta = \min_m [E_m - E_0] \approx c\xi^{-1}. \tag{12.39}$$

If Δ does not vanish with the size of the system, it is a thermodynamically meaningful gap in the excitation spectrum called *Haldane's gap*.

In the absence of topological Berry phases Υ, the ground state of the quantum antiferromagnet is described by the classical energy of the $d+1$ dimensional NLSM, where f is the coupling constant of the classical problem. At finite (but low) temperature β^{-1}, the quantum antiferromagnet is mapped onto the NLSM on a *slab* of finite width $c\beta$ in the imaginary time dimension. In Chapters 13 and 14, we study the NLSM and derive its long-range correlations. In Chapter 15, we discuss the correlations in the quantum Heisenberg antiferromagnet with the effects of the Berry phases.

One must be cautioned that Haldane's mapping is valid under the restrictive condition (12.10), which can be achieved in the limit of $S \gg 1$.

TABLE 12.1. The Nonlinear sigma model parameters for the nearest neighbor Heisenberg antiferromagnet.

parameter	value
Λ	a^{-1}
ρ_s	$JS^2 a^{2-d}$
χ_0	$(4dJa^d)^{-1}$
c	$2JSa\ d^{-\frac{1}{2}}$
f	$2\sqrt{d}\ S^{-1}$

Thus the large S (semiclassical) limit corresponds to the *weak coupling* limit of the NLSM, $f \ll 1$.

As a concrete example, we list in Table 12.1 the NLSM parameters for the simplest nearest neighbor Heisenberg antiferromagnet on a cubic lattice. These are given by the definitions and equations of this chapter.

12.5 Exercises

1. Following the derivation of the contiunuum approximation for Z (12.32), show that the Green's function (10.21) can be approximated by the *real-time* path integral of the NLSM field theory:

$$G(t) \propto \int_\Lambda \mathcal{D}\hat{n}\ \delta(|\hat{n}| - 1)\ e^{i\Upsilon}\ \exp\left(-i \int_0^t dt' \int dx^d \mathcal{L}_{NLSM}^{d+1}\right),$$

$$\mathcal{L}_{NLSM}^{d+1} = \frac{\Lambda^{d-1}}{2f} \sum_{\mu=1}^{d+1} \partial_\mu \hat{n} \cdot \partial^\mu \hat{n}, \tag{12.40}$$

where $x_{d+1} = ict$, and $x_\mu x^\mu = \mathbf{x} \cdot \mathbf{x} - x_{d+1}^2$ is the *Minkowski product*.

2. Derive the classical dynamics for the field $\hat{n}(\mathbf{x}, t)$ by varying the total action in (12.40) with respect to $\hat{n}$. Show that the equation of motion is

$$\partial_\mu \partial^\mu \hat{n} = 0. \tag{12.41}$$

Linearize (12.41) for small fluctuations about $\hat{n} = \hat{x}$, and solve for the eigenmodes (spin waves). Relate these spin wave modes to the polarizations and dispersions of Heisenberg spins.

Bibliography

Much of this chapter follows Haldane's lecture notes:

- F.D.M. Haldane, *Two-Dimensional Strongly Correlated Electron Systems* edited by Z.Z. Gan and Z.B. Su (Gordon and Breach, 1988), p. 249.

The original derivations of Haldane's mapping used Holstein–Primakoff operators and semiclassical equations of motion; see

- F.D.M. Haldane, Phys. Lett. **A93**, 464 (1983);
- F.D.M. Haldane, Phys. Rev. Lett. **50**, 1153 (1983).

13

Nonlinear Sigma Model: Weak Coupling

13.1 The Lattice Regularization

In the previous chapter, the Nonlinear Sigma Model (NLSM) in d dimensions has emerged as the continuum approximation to the quantum Heisenberg antiferromagnet in $d-1$ dimensions with additional Berry phases. The partition function is

$$Z_{NLSM} = \int_{\Lambda} \mathcal{D}\hat{\mathbf{n}} \, \exp\left(-\frac{\Lambda^{d-2}}{2f}\int d^d x \sum_{\mu=1}^{d} \partial_\mu \hat{\mathbf{n}} \cdot \partial_\mu \hat{\mathbf{n}}\right). \qquad (13.1)$$

$|\hat{\mathbf{n}}(\mathbf{x})| = 1$ is a unit vector and $\mathbf{x} \in R^d$. Λ is the momentum cutoff, i.e., the shortest wavelength included in the $\mathcal{D}\hat{\mathbf{n}}$. The path integral (13.1) is not well defined until we specify a regularization procedure. We consider a classical Heisenberg model on a d-dimensional cubic lattice, with $\mathcal{N}$ sites and lattice constant a. Its partition function is

$$Z_{HM} = \int \prod_{i=1}^{\mathcal{N}} d\hat{\Omega}_i \, \exp\left(\frac{\bar{J}}{T}\sum_{\langle ij\rangle} \hat{\Omega}_j \cdot \hat{\Omega}_j\right). \qquad (13.2)$$

The continuum limit of Z_{HM} is given by the substitutions

$$\hat{\Omega}_i \to \eta_i \hat{\mathbf{n}}(\mathbf{x}_i), \qquad (13.3)$$

where η_i is 1 ($e^{i\mathbf{X}_i \vec{\pi}}$) for the ferromagnet (antiferromagnet). Sums and differences are substituted by

$$\sum_i F(\mathbf{x}_i) \to a^{-d}\int d^d x \, F(\mathbf{x}),$$

$$\hat{\Omega}(\mathbf{x}_i + a\hat{x}_\mu) - \hat{\Omega}(\mathbf{x}_i) \to a\partial_\mu \hat{\mathbf{n}}, \qquad (13.4)$$

and a dimensionless coupling constant is given by

$$\frac{T}{\bar{J}} \leftrightarrow f. \qquad (13.5)$$

The measure is replaced by

$$\mathcal{D}\hat{\Omega} \rightarrow \prod_{|\mathbf{q}|\leq\Lambda} d\hat{n}_{\mathbf{q}}, \tag{13.6}$$

where Λ is the radius of a spherical (sph) Brillouin zone that has the same number of degrees of freedom as the cubic zone ($cube$) of the lattice

$$\mathcal{N}^{-1} \sum_{|k_\mu|\leq\frac{\pi}{a}}^{cube} = \mathcal{N}^{-1} \sum_{|k|\leq\Lambda}^{sph} = 1. \tag{13.7}$$

This determines the spherical Brillouin zone's radius to be

$$\Lambda = \begin{cases} \pi/a & d=1 \\ 2\sqrt{\pi}/a & d=2 \\ (6\pi^2)^{1/3}/a & d=3 \end{cases}. \tag{13.8}$$

Expanding the Hamiltonian in (13.2) in terms of gradients of $\hat{n}$ and keeping up to quadratic terms, the relation

$$Z_{HM}(T/\bar{J}, a) \approx Z_{NLSM}(f, \Lambda^{-1}) \, e^{d\mathcal{N}/f} \tag{13.9}$$

is established at low temperatures ($f << 1$). The factor $e^{d\mathcal{N}/f}$ is the Boltzmann weight of the classical ground state. The continuum approximation holds also for the generating functional (see Appendix B) and for the long-wavelength spin correlations, as long as the spin correlation length ξ is much larger than Λ^{-1}. In this regime, the NLSM properties are weakly dependent on the regularization procedure, and therefore it serves as a model for diverse Heisenberg models. The crucial features of these models are their $O(3)$ symmetry of the ground state manifold and their short-range interactions. Other details will primarily affect the choice of f and Λ.

The "nonlinearity" of the NLSM is due to the unimodular constraint $|\hat{n}(\mathbf{x})| = 1$. One can write, say the first component, in terms of the other components as

$$n^{(1)}(\mathbf{x}) = \sqrt{1 - \left[n^{(2)}(\mathbf{x})\right]^2 - \left[n^{(3)}(\mathbf{x})\right]^2}. \tag{13.10}$$

The measure is then replaced by

$$\int \mathcal{D}\hat{n} = \int \mathcal{D}n^{(2)} \mathcal{D}n^{(3)} \frac{\Theta\left[1 - (n^{(2)})^2 - (n^{(3)})^2\right]}{\sqrt{1 - (n^{(2)})^2 - (n^{(3)})^2}}. \tag{13.11}$$

The constraint introduces anharmonic interactions between the field components. Thus, we can expect that the correlations of the NLSM may differ from a free (Gaussian) model, where the measure is $\mathcal{D}n^{(2)}\mathcal{D}n^{(3)}$. An important effect of the unimodular constraint is the existence of topologically stable excitations ("*solitons*") in two dimensions.[1]

---

[1] See Skyrme, and Belavin and Polyakov.

As a continuum limit of the classical Heisenberg model, the $O(3)$ NLSM must obey the Mermin and Wagner theorem, which was proven in Section 6.2. That theorem forbids spontaneously broken symmetry at any finite temperature (or $f > 0$) in $d = 1, 2$.

In the next section, we shall demonstrate how the $O(n)$ NLSM is disordered by fluctuations of the order parameter using a weak coupling expansion in f.

13.2 Weak Coupling Expansion

First, we generalize the NLSM from $O(3)$ to $O(n)$ symmetry. Since the following calculations do not entail any extra work for general n, we might as well keep n arbitrary. We define n orthonormal basis vectors:

$$\hat{\mathbf{e}}^{\alpha}, \qquad \alpha = 0, \ldots, n-1 ,$$
$$n^{\alpha} = \hat{\mathbf{n}} \cdot \hat{\mathbf{e}}^{\alpha}. \tag{13.12}$$

The partition function is

$$Z_{O(n)} = \int \mathcal{D}\hat{\mathbf{n}} \, \exp\left[-\frac{\Lambda^{d-2}}{2f} \int d^d x \sum_{\mu\alpha} (\partial_\mu n^\alpha)^2\right] ,$$
$$\mathcal{D}\hat{\mathbf{n}} = \prod_{\alpha} \mathcal{D}n^\alpha \, \delta\left[1 - \sum_{\alpha=0}^{n-1} (n^\alpha)^2\right]. \tag{13.13}$$

At low temperatures, i.e., weak coupling $f \ll 1$, we expect the dominant configurations to be close to an ordered ground state. This requires us to break the $O(n)$ symmetry of the path integral and choose a particular uniform ground state configuration, say

$$\bar{\hat{\mathbf{n}}} = \hat{\mathbf{e}}^0. \tag{13.14}$$

The small fluctuations, $|\hat{\mathbf{n}} - \hat{\mathbf{e}}^0| \ll 1$, are parametrized by

$$\hat{\mathbf{n}} = \hat{\mathbf{e}}_0 \sqrt{1 - \bar{\phi}^2} + \sum_{a=1}^{n-1} \phi^a \hat{\mathbf{e}}^a, \tag{13.15}$$

where

$$\bar{\phi}^2 \equiv \sum_{a=1}^{n-1} (\phi^a)^2 . \tag{13.16}$$

Here and henceforth we use Latin indices (e.g., a, b) to denote transverse flucuations which exclude the longitudinal $\hat{\mathbf{e}}_0$ direction.

The expansion of the NLSM to leading order in $|\delta\hat{n}|$ yields

$$Z \approx \int \mathcal{D}\phi \exp\left(-\mathcal{S}^{(2)}[\phi]\right)\left[1 + \mathcal{O}(f)\right] ,$$

$$\mathcal{S}^{(2)} = \frac{\Lambda^{d-2}}{2f}\int d^d x \sum_{a,\mu}(\partial_\mu\phi^a)^2 + \mathcal{O}(\phi^3)$$

$$= \frac{1}{2f\Lambda^2\mathcal{N}}\sum_{\mathbf{k},a}\mathbf{k}^2\phi_{\mathbf{k}}^a\phi_{-\mathbf{k}}^a, \tag{13.17}$$

where

$$\phi_{\mathbf{k}}^a = \Lambda^d\int d^d x\, e^{-i\mathbf{k}\cdot\mathbf{x}}\phi^a(\mathbf{x}). \tag{13.18}$$

$\mathcal{S}^{(2)}$ is the harmonic or "spin wave" energy functional. Using (13.15) and (13.17), we can calculate the spin wave contribution to the local fluctuations:

$$\langle|\delta\hat{n}(0)|^2\rangle \approx \frac{1}{Z\mathcal{N}}\sum_{\mathbf{k}a}\int\mathcal{D}\hat{n}\,\phi_{\mathbf{k}}^a\phi_{-\mathbf{k}}^a\exp\left(-\mathcal{S}^{(2)}[\phi]\right)$$

$$= \frac{(n-1)\Lambda^2 f}{(2\pi)^d}\lim_{\tilde{\Lambda}\to 0}\int_{\tilde{\Lambda}}^{\Lambda}\frac{d^d k}{|\mathbf{k}|^2}$$

$$\sim \begin{cases}(n-1)f\lim_{\tilde{\Lambda}\to 0}\tilde{\Lambda}^{-1} = \infty & d=1 \\ -\frac{(n-1)f}{2\pi}\lim_{\tilde{\Lambda}\to 0}\ln\tilde{\Lambda} = \infty & d=2 \\ \frac{(n-1)f\Lambda}{2\pi^2} < \infty & d=3\end{cases} \tag{13.19}$$

The spin wave fluctuations diverge for $n \geq 2$ in one and two dimensions for all $f > 0$. Thus, naive weak coupling expansion fails since the assumption of spontaneously broken symmetry is invalid. This we have known all along by Mermin and Wagner's Theorem 6.2. By Haldane's mapping, the *ground state* of the quantum Heisenberg antiferromagnet in one dimension should be disordered (a spin liquid) for all finite spin sizes $S < \infty$.

We have used the term "naive expansion" to hint that more sophisticated methods can treat the infrared divergences in (13.19), as we shall see in the following section.

13.3   Poor Man's Renormalization

The infrared divergences in (13.19) indicate that it is necessary to abandon the assumption of broken symmetry and to preserve the $O(n)$ symmetry of the path integral. It is still possible, however, to use the smallness of f for approximating the correlations in the disordered phase. This will be done by expanding the integrand about *slowly varying* configurations $\hat{n}^0(\mathbf{x})$, which still need to be integrated over. A sequential elimination of the modes of large momenta (the "fast modes") renormalizes the energy functional of the remaining "slow" modes. In this section, we derive the renormalization group transformation which is embodied in the β function. In the following section, we use the β function to determine the correlation length.

Following Polyakov, we separate $\hat{\mathbf{n}}$ into its *slow* and *fast* degrees of freedom, $\hat{\mathbf{n}}^0$ and ϕ^a, respectively:

$$\hat{\mathbf{n}}(\mathbf{x}) = \hat{\mathbf{n}}^0(\mathbf{x})\sqrt{1 - \bar{\phi}^2} + \sum_{a=1}^{n-1} \phi^a \hat{\mathbf{e}}^a(\mathbf{x}),$$

$$\bar{\phi}^2 = \sum_{a=1}^{n-1}(\phi^a)^2, \tag{13.20}$$

where a comoving basis is defined as

$$\hat{\mathbf{e}}^0(\mathbf{x}) \equiv \hat{\mathbf{n}}^0(\mathbf{x}),$$
$$\hat{\mathbf{e}}^\alpha(\mathbf{x}) \cdot \hat{\mathbf{e}}^\beta(\mathbf{x}) = \delta_{\alpha\beta}, \quad \alpha = 0, 1, \ldots, n-1. \tag{13.21}$$

Greek indices include the longitudinal direction $\hat{\mathbf{n}}^0$, while Latin indices are restricted to the transverse directions $a, b = 1, \ldots, n-1$.

The relation between the coordinate systems at nearby points in space defines the *Gauge potentials*[2]

$$\tilde{A}_\mu^{\alpha\beta} \equiv \hat{\mathbf{e}}^\alpha \cdot \partial_\mu \hat{\mathbf{e}}^\beta, \tag{13.22}$$

which describe the variations of the moving orthonormal basis by

$$\partial_\mu \hat{\mathbf{n}}^0 = \sum_a \tilde{A}_\mu^{a0} \hat{\mathbf{e}}^a,$$

$$\partial_\mu \hat{\mathbf{e}}^a = \sum_b (\tilde{A}_\mu^{ba} \hat{\mathbf{e}}^b) - \tilde{A}_\mu^{a0} \hat{\mathbf{n}}^0. \tag{13.23}$$

Since $\hat{\mathbf{e}}^\alpha$ are orthonormal, $\tilde{A}_\mu^{\alpha\beta}$ must be antisymmetric:

$$\tilde{A}_\mu^{\alpha\beta} = -\tilde{A}_\mu^{\beta\alpha}. \tag{13.24}$$

We note a useful identity which will be used shortly:

$$\sum_a (\tilde{A}_\mu^{a0})^2 = (\partial_\mu.\hat{\mathbf{n}}^0)^2. \tag{13.25}$$

An intermediate momentum scale $\tilde{\Lambda}$ is chosen,

$$0 < \tilde{\Lambda} << \Lambda, \tag{13.26}$$

---

[2]In differential geometry they are called *connections*.

which separates between the "fast" fields, i.e.,

$$\phi^a(x) \equiv \sum_{\tilde{\Lambda} \le |\mathbf{k}| \le \bar{\Lambda}} \exp[i\mathbf{k}x] \, \phi^a_{\mathbf{k}}, \tag{13.27}$$

and the slow fields

$$\mathbf{X}(x) = \sum_{0 \le |\mathbf{k}| < \tilde{\Lambda}} \exp[i\mathbf{k}x] \, \mathbf{X}_{\mathbf{k}} \,, \qquad \mathbf{X} = \hat{\mathbf{e}}^\alpha, \tilde{A}^{\alpha,\beta}_\mu. \tag{13.28}$$

Inserting (13.20) into the NLSM Lagrangian, we obtain

$$\begin{aligned}
\mathcal{L} &= \frac{\Lambda^{d-2}}{2f} \int d^d x \sum_{\mu=1,d} \partial_\mu \hat{\mathbf{n}} \cdot \partial_\mu \hat{\mathbf{n}} \\
&= \frac{\Lambda^{d-2}}{2f} \sum_{\mu ab} \left[\left(\partial_\mu \sqrt{1-\bar{\phi}^2} - \tilde{A}^{a0}_\mu \phi^a \right)^2 \right. \\
&\qquad \left. + \left(\partial_\mu \phi^a + \tilde{A}^{ab}_\mu \phi^b + \tilde{A}^{a0}_\mu \sqrt{1-\bar{\phi}^2} \right)^2 \right].
\end{aligned} \tag{13.29}$$

We expand the Lagrangian[3] to quadratic order in ϕ, which yields

$$\begin{aligned}
Z &\approx \int_{\tilde{\Lambda}} \mathcal{D}\hat{\mathbf{n}}^0 \, \exp\left(-\int_{\tilde{\Lambda}} d^d x \, \mathcal{L}[\hat{\mathbf{n}}^0] \right) Z^{(2)}[\hat{\mathbf{n}}^0] \, \left[1 + \mathcal{O}(\phi^2, f^{-1}\phi^3) \right], \\
Z^{(2)} &= \int_\Lambda \mathcal{D}\phi \, \exp\left(-\int_\Lambda d^d x \, \mathcal{L}^{(2)} \left[\hat{\mathbf{n}}^0, \phi \right] \right),
\end{aligned} \tag{13.30}$$

where relation (13.25) yields the term $\mathcal{L}[\hat{\mathbf{n}}^0]$ and the fast ("spin wave") Lagrangian is given by

$$\mathcal{L}^{(2)} = \frac{\Lambda^{d-2}}{2f} \left[\sum_{\mu ab} \left(\partial_\mu \phi^a + \tilde{A}^{ab}_\mu \phi^b \right)^2 + \tilde{A}^{a0}_\mu \tilde{A}^{b0}_\mu \left(\phi^a \phi^b - \delta_{ab} \bar{\phi}^2 \right) \right]. \tag{13.31}$$

We neglect the cubic and higher powers of ϕ and confine the present analysis to the leading-order terms in f. In the renormalization group jargon, we are performing a *"one loop level"* calculation.

In (13.31), we have also neglected the cross terms

$$\tilde{A}^{ab}_{\mathbf{q}} \phi^b_{\mathbf{k}-\mathbf{q}} \tilde{A}^{a0}_{-\mathbf{k}}, \tag{13.32}$$

which couple slow and fast fields. By choosing $\tilde{\Lambda} \ll \Lambda$, we make sure that the contribution of such terms is small since they vanish for all $\mathbf{k} > 2\tilde{\Lambda}$, i.e., for most of the Brillouin zone.

---

[3]The Jacobian of transformation (13.20) does not contribute to the quadratic corrections.

$\mathcal{L}^{(2)}$ describes the high momenta fluctuations about $\hat{\mathbf{n}}^0$. The alert reader will notice that for $n > 2$ there is a certain degree of arbitrariness in the choice of the transverse unit vectors $\hat{\mathbf{e}}^a$ in (13.20). For $n > 2$, one can continuously rotate the transverse coordinate system in the $n - 1$ dimensional subspace at each point in space using any orthogonal $(n - 1) \times (n - 1)$ matrix R such that

$$\phi^a \rightarrow R^{ab}\phi^b, \quad R^T R = 1, \tag{13.33}$$

which transforms the covariant derivative in the first term of (13.31) as

$$\partial_\mu \delta_{ab} + \tilde{A}_\mu^{ab} \rightarrow \partial_\mu \delta_{ab} + \tilde{A}_\mu^{ab} + [(\partial_\mu R)R^{-1}]^{ab}. \tag{13.34}$$

The invariance of $\mathcal{L}^{(2)}$ under (13.34) is a *gauge symmetry* of the fast modes Lagrangian. It is easy to see that $\mathcal{L}^{(2)}$ can only depend on derivatives of $\tilde{A}_\mu^{ab}$. Since these potentials are slow fields, these terms are higher-order derivatives of $\hat{\mathbf{n}}^0$ than the leading term $\mathcal{L}[\hat{\mathbf{n}}^0]$. Therefore, for the following discussion we shall neglect the gauge potentials in $\mathcal{L}^{(2)}$. In the Exercises, we shall see that the gauge potentials are important when expanding about a slow field with finite topological density.

By performing the Gaussian integral in (13.30), we obtain

$$Z^{(2)}[\hat{\mathbf{n}}^0] \propto \exp\left[-\frac{1}{2}\text{Tr}\ln\left(\Pi_0 - \Pi_1\right)\right]. \tag{13.35}$$

The overall normalization constants are independent on $\hat{\mathbf{n}}^0$, and unimportant for the correlation functions. Π_0 and Π_1 are matrices in the fast mode space $|\mathbf{k}| > \tilde{\Lambda}$:

$$(\Pi_0)_{\mathbf{k},\mathbf{k}'} = -(\partial_\mu^2)\delta_{ab} = k^2\,\delta_{\mathbf{k},-\mathbf{k}'}\delta_{ab}\,,$$

$$(-\Pi_1)_{\mathbf{k},\mathbf{k}'} = \delta_{\mathbf{k},-\mathbf{k}'}\left[\tilde{A}_\mu^{a0}\tilde{A}_\mu^{b0} - \delta_{ab}\sum_c(\tilde{A}_\mu^{c0})^2\right]. \tag{13.36}$$

We can expand the exponential in $Z^{(2)}$ in powers of $\tilde{A}^{a0}$ using the identity

$$-\frac{1}{2}\text{Tr}\ln\left(\Pi_0 - \Pi_1\right) = -\frac{1}{2}\text{Tr}\ln\Pi_0 + \frac{1}{2}\sum_{n=1}^\infty \frac{1}{n}\text{Tr}\left(\Pi_0^{-1}\Pi_1\right)^n. \tag{13.37}$$

It is consistent with our earlier approximation to neglect higher derivatives and thus we may keep only up to the $n = 1$ term, which is

$$\frac{1}{2}\text{Tr}\left(\Pi_0^{-1}\Pi_1\right) = -\frac{1}{2N}\sum_{\tilde{\Lambda}<\mathbf{k}<\Lambda}\frac{1}{k^2}\sum_{\mu a}\left[(\tilde{A}_\mu^{a0})^2 - (n-1)(\tilde{A}_\mu^{a0})^2\right]$$

$$\approx \frac{\Lambda^{d-2}\Delta_d}{2}\sum_\mu(\partial_\mu\hat{\mathbf{n}}^0)^2, \tag{13.38}$$

where we have used (13.25), and defined the dimensionless constant

$$
\Delta_d(\tilde{\Lambda}/\Lambda) \;=\; (n-2)\Lambda^{2-d} \int_{\tilde{\Lambda}<|\mathbf{k}|<\Lambda} \frac{d^d k}{(2\pi)^d} \frac{1}{k^2}
$$

$$
=\; \begin{cases}
\frac{(n-2)}{\zeta_1}\left[(\tilde{\Lambda}/\Lambda)^{d-2}-1\right] & d=1 \\[2mm]
-\frac{n-2}{\zeta_2}\,\ln(\tilde{\Lambda}/\Lambda) & d=2 \,, \\[2mm]
\frac{(n-2)}{\zeta_3}\left[1-(\tilde{\Lambda}/\Lambda)^{d-2}\right] & d=3
\end{cases}
\qquad (13.39)
$$

where $\zeta_1 = \pi$, $\zeta_2 = 2\pi$, and $\zeta_3 = 2\pi^2$.

By (13.38), we find that the correction to the NLSM from integrating out the fast modes is proportional to the NLSM Lagrangian itself! By incorporating the corrections into the NLSM and replacing the cutoff $\Lambda \to \tilde{\Lambda}$, we obtain

$$
Z \;\approx\; \int_{\tilde{\Lambda}} \mathcal{D}\hat{\mathbf{n}}^0 \; \exp\left[-\frac{\tilde{\Lambda}^{d-2}}{2\tilde{f}} \int_{\tilde{\Lambda}} d^d x (\partial_\mu \hat{\mathbf{n}}^0)^2 \right],
\qquad (13.40)
$$

where the renormalized coupling constant is

$$
\tilde{f} \;=\; (\tilde{\Lambda}/\Lambda)^{d-2} \frac{f}{1 - f\Delta_d(\tilde{\Lambda}/\Lambda)}.
\qquad (13.41)
$$

Equation (13.41) is valid deep in the perturbative regime $f, \tilde{f} \ll 1$. Otherwise, it is not justified to neglect the cubic and higher terms in the Lagrangian as we have done in (13.30).

For $d > 2$, Δ does not diverge in the infrared limit $\tilde{\Lambda} \to 0$, and the correction to $\tilde{f}$ is small. On the other hand, for $d \le 2$ and $n \ge 3$, Δ_d diverges in the infrared limit.

A comparison of (13.19) to (13.39) is instructive. Although for $d=1,2$, the ground state fluctuations diverge as $n-1$, the infrared divergences of Δ_d in (13.39) occur only for $n \ge 3$. Thus, we see that the $n=2$ case is special. For $n=2$ there is only one transverse spin wave mode, and to first loop order, the spin wave theory is effectively noninteracting. On the other hand, for $n>2$ the interactions between spin wave modes flow under renormalization to the strong coupling regime.

The $O(2)$ NLSM is the continuum limit of the xy model

$$
\mathcal{H}_{xy} \;=\; \bar{J} \sum_{\langle ij \rangle} (\hat{\Omega}_i^x \hat{\Omega}_j^x + \hat{\Omega}_i^y \hat{\Omega}_j^y) \;\rightsquigarrow\; \mathcal{L}_{O(2)}.
\qquad (13.42)
$$

For $d=2$, the β function of $\mathcal{L}_{O(2)}$ is governed by the higher-order (cubic) power of f, which was not calculated above. It turns out that the continuum theory has a weak coupling phase, with "quasi-long-range order" for $f > 0$, i.e., power law decaying correlations. At higher temperatures (larger f), the continuum approximation (13.42) breaks down, and vortex singularities become important. Kosterlitz and Thouless (KT) have shown that unbinding of vortex pairs destroys the quasi-long-range order at the *Kosterlitz–Thouless* temperature, whereupon the spin correlations decay exponentially at long distances. $O(2)$ models and the KT transition are very important in the context of many physical phenomena, including superfluidity and superconductivity. They are reviewed at length in the literature. Here we shall confine our further discussions to Heisenberg-type models with $n \geq 3$.

In summary, to leading order in f, the NLSM scales onto itself under a change of the upper cutoff, and we have obtained (approximately) the renormalization equation for the coupling constant. Our analysis assumes that the only coupling constant that grows as we renormalize the model is f, i.e., the higher-order interactions, and higher derivatives terms, are either *"irrelevant"* or *"marginal."* This assumption should be verified by an explicit calculation. Here we blissfully ignore this issue, which is why this type of analysis is called *"Poor Man's Scaling"*.[4] A more careful treatment of the generating functional using a $d-2$ expansion was carried out by Brezin and Zinn-Justin.

13.4 The β Function

Here we drop the tilde off $\tilde{f}$, and denote the cutoff dependent coupling constant by $f(\Lambda)$. In (13.41) we have found the leading-order renormalization of f under changing $\Lambda \rightarrow \tilde{\Lambda}$. The β function, which governs the "flow" of the coupling constant under renormalization, is defined as

$$\frac{df}{d\ln(\tilde{\Lambda}/\Lambda)} \equiv \beta, \tag{13.43}$$

which by ((13.41) yields

$$\beta = (d-2)f - \frac{n-2}{\zeta_d}f^2. \tag{13.44}$$

Thus β depends *explicitly* only on f and not on Λ or $\tilde{\Lambda}$. This property is called *"universality."* The zeros of the β function are *"fixed points"* of the renormalization transformation:

$$\beta(f_c) = 0 \ . \tag{13.45}$$

---

[4]The term was coined by P.W. Anderson in reference to Anderson, Yuval, and Hamman's renormalization of the Kondo model.

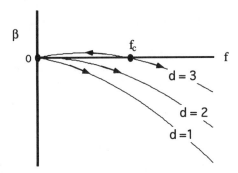

FIGURE 13.1. The β function and the flow of f as Λ is reduced.

For $\beta < 0$ the coupling constant decreases and for $\beta > 0$ it increases under downward renormalization of $\tilde{\Lambda}$. The fixed points are:

$$
\begin{aligned}
d = 1 \quad &: \quad f_c = 0 \quad \text{unstable}\,, \\
d = 2 \quad &: \quad f_c = 0 \quad \text{unstable}\,, \\
d = 3 \quad &: \quad \begin{cases} f_c = 0 & \text{stable} \\ f_c = \frac{2\pi^2}{n-2} & \text{unstable} \end{cases}
\end{aligned}
\tag{13.46}
$$

where "stable" and "unstable" refer to the flow of $f \approx f_c$ under renormalization, whether toward or away from the fixed point. As shown in Fig. 13.1, in one and two dimensions the coupling constant flows to strong coupling. In three dimensions, there is a weak coupling phase and a strong coupling phase. For $f < f_c$, the coupling flows to the weakly interacting fixed point. This implies that perturbation theory works well and low-order spin wave theory about the ordered ground state is valid. As shown in Fig. 13.1, for $f > f_c$, the system flows to strong coupling, which is disordered. f_c is the critical coupling constant which separates between the broken symmetry and disordered phases.

Deep in the strong coupling phase, we assume that the momentum dependent structure factor is smooth, and the short-range order is manifested by a maximum at $\mathbf{q} = 0$, parametrized by

$$
\frac{1}{\mathcal{N}} \langle \hat{n}_{\mathbf{q}} \hat{n}_{-\mathbf{q}} \rangle \approx \frac{C}{|\mathbf{q}|^2 + \xi^{-2} + \mathcal{O}(|\mathbf{q}^3|)},
\tag{13.47}
$$

where C is a dimensional constant. The Fourier transform of (13.47) is the "*Ornstein–Zernicke*" correlation function. Its asymptotic form at large

$|\mathbf{x}|/\xi$ is[5]

$$\langle \hat{\mathbf{n}}(0)\hat{\mathbf{n}}(\mathbf{x})\rangle \propto |\mathbf{x}|^{-(d-1)/2} \exp(-|\mathbf{x}|/\xi)\left(1 + \frac{\xi(d-3)}{|\mathbf{x}|}\right), \tag{13.48}$$

where the correlation length ξ governs the exponential decay of the correlations at large distances. ξ is a characteristic length scale of the correlations. It should not matter whether we evaluate ξ using the NLSM with the high cutoff Λ and bare coupling constant $f(\Lambda)$ or the renormalized model with $\tilde{\Lambda}$ and $f(\tilde{\Lambda})$. This invariance can be phrased as a functional identity:

$$\xi[f(\Lambda), \Lambda] = \xi[f(\tilde{\Lambda}), \tilde{\Lambda}]. \tag{13.49}$$

The derivative of the left-hand side with respect to $\ln(\tilde{\Lambda})$ vanishes. Therefore, we obtain

$$\frac{\partial \xi}{\partial \ln(\tilde{\Lambda})} + \beta \frac{\partial \xi}{\partial f} = 0. \tag{13.50}$$

Now, since ξ has dimensions of distance and f is dimensionless, by dimensional analysis we know that

$$\xi(\tilde{\Lambda}, f) = \tilde{\Lambda}^{-1}\phi(f), \tag{13.51}$$

where ϕ is a dimensionless function. Thus,

$$\frac{\partial \xi}{\partial \ln(\tilde{\Lambda})} = -\xi, \tag{13.52}$$

whereupon (13.50) has the explicit solution

$$\int_{\xi_0}^{\xi} \frac{d\xi}{\xi} = \int_{f_0}^{f} \frac{df}{\beta(f)},$$
$$\rightarrow \xi = \xi_0 \exp\left(\int_{f_0}^{f} \frac{df}{\beta(f)}\right). \tag{13.53}$$

ξ_0 and f_0 are constants of integration which can be fixed by knowing the correlation length at strong coupling, as discussed shortly.

In three dimensions, β vanishes linearly at f_c, and thus ξ diverges as a power law,

$$\xi^{d=3} \sim \left(\frac{f - f_c}{f_0}\right)^{1/\beta'(f_c)},$$
$$\beta'(f_c) = -1 + \mathcal{O}(1). \tag{13.54}$$

---

[5]See Fisher's paper in the bibliography.

In two dimensions, by (13.44) the β function vanishes quadratically at $f_c = 0$. We can integrate the right-hand side of (13.53) from the strong coupling regime where $f_0 \gg 1$, and the correlation length is $\xi_0 = O(a)$, to the physical value of f:

$$\xi^{d=2} \sim \xi_0 \exp\left(\frac{2\pi}{(n-2)f}\right). \tag{13.55}$$

In order to determine ξ_0 for a given lattice model, one can fit (13.55) to a numerically determined correlation length at strong coupling (i.e., high temperatures). ξ_0/a is a numerical constant which will depend on the short-range details of the lattice model.[6]

In one dimension (13.53) integrates to yield

$$\xi^{d=1} \sim \frac{\xi_0 f_0}{f}, \tag{13.56}$$

where $\xi_0 f_0$ is of the order of Λ^{-1}.

13.5   Exercises

1. For the $O(3)$ model, write the two spin wave coordinates as a single complex field:

$$\psi(\mathbf{x}) = \phi^1(\mathbf{x}) + i\phi^2(\mathbf{x}), \tag{13.57}$$

and define a $U(1)$ gauge field as

$$A_\mu[\hat{\mathbf{n}}^0] = -\frac{i}{2}\tilde{A}^{12}[\hat{\mathbf{n}}^0]. \tag{13.58}$$

Find the spin wave eigenvalues ϵ_n in the presence of finite Gauge potentials by diagonalizing the quadratic form in $\mathcal{L}^{(2)}$ of (13.31):

$$\int d^d x \left(\partial_\mu \phi^a + \tilde{A}_\mu^{ab}\phi^b\right)^2. \tag{13.59}$$

Show that ϵ_n can be obtained from the equation

$$(\partial_\mu - i2A_\mu)^2 \psi_n = \epsilon_n \psi_n. \tag{13.60}$$

Note: (13.60) describes a free boson of charge 2 in the presence of an electromagnetic vector potential A_μ.

2. Prove that

$$(\partial_\mu A_\nu - \partial_\nu A_\mu) = \frac{1}{2}\partial_\mu \hat{\mathbf{n}} \times \partial_\nu \hat{\mathbf{n}} \cdot \hat{\mathbf{n}} \tag{13.61}$$

for the $O(3)$ NLSM gauge field $A(\hat{\mathbf{n}})$ as defined in (13.58).

---

[6]See Shenker and Tobochnik.

3. For a two-dimensional system, assume a uniform topological density $B = \partial_x \hat{n} \times \partial_y \hat{n} \cdot \hat{n}$, such as found near the center of a large Skyrmion configuration. Choose a convenient gauge to express A_x, A_y, and solve for the eigenvalues of (13.60). *Hint: Use the theory of Landau levels of a particle in a uniform magnetic field.*[7]

Bibliography

The poor man's renormalization follows Polyakov's original derivations in

- A.M. Polyakov, Phys. Lett. B **59**, 79 (1975);
- A.M. Polyakov, *Gauge Fields and Strings* (Harwood, 1987), Chapter 2.

The d–2 expansion for the nonlinear sigma model to two loop order was calculated by

- E. Brézin and J. Zinn-Justin, Phys. Rev. B **14**, 3110 (1976).

The correlation length of the nonlinear sigma model was carried out using Monte Carlo simulations on finite lattices by

- S.H. Shenker and J. Tobochnik, Phys. Rev. B **22**, 4462 (1980).

The Fourier transformation for the Ornstein–Zernicke correlation function is evaluated in

- M.E. Fisher, Physica **28**, 172 (1962).

Solitons in the two-dimensional NLSM, or Skyrmions, were found by

- T. Skyrme, Proc. R. Soc. London Ser. A **260**, 127 (1961);
- A.A. Belavin and A.M. Polyakov, JETP Lett. **22**, 245 (1975).

On the effects of the gauge potential A^μ_{ab} on the spin wave spectrum, see

- A. Auerbach, B. Larson, and G. Murthy, Phys. Rev. B **43**, 11515 (1991).

The two-dimensional xy model and the effects of vortices were elucidated by

- J.M. Kosterlitz and D.J. Thouless, J. Phys. C **6**, 1181 (1973);
- V.L. Berezinskii, Sov. Phys. JETP **34**, 610 (1972).

---

[7]See L.D. Landau and E.M. Lifshits *Quantum Mechanics* (Pergamon Press), p. 457, or see Auerbach, Larson, and Murthy.

14

The Nonlinear Sigma Model: Large N

Chapter 13 solved for the correlations of the d-dimensional Nonlinear Sigma Model (NLSM) in the weak coupling $f << 1$ regime. The weak coupling regime coincides with the low-temperature limit of the classical Heisenberg model on a d-dimensional lattice. By Haldane's mapping (see Chapter 12) it is also equivalent to the large S limit of the $d-1$ dimensional quantum Heisenberg antiferromagnet, i.e., the *semiclassical* limit. In this chapter, we study an alternative approach to the NLSM. We introduce the *complex projective* representations called the CP^{N-1} *models*, where N is the number of complex fields. The physical $O(3)$ model is described by $N = 2$. Basically, the large N limit is given by a saddle point of the path integral, i.e., a mean field theory. This mean field theory recovers the correlation length ξ obtained earlier by poor man's scaling in Section 13.4.

14.1 The CP^1 Formulation

The $O(3)$ NLSM can be represented by two complex (spinor) fields

$$\mathbf{z}(\mathbf{x}) = [z_1(\mathbf{x}), z_2(\mathbf{x})] \tag{14.1}$$

such that the components of $\hat{\mathbf{n}}$ are given by the bilinear forms

$$n^\alpha(\mathbf{x}) = \mathbf{z}^\dagger(\mathbf{x})\sigma^\alpha\mathbf{z}(\mathbf{x}) , \quad \alpha = x, y, z, \tag{14.2}$$

where σ^α are 2×2 Pauli matrices (see (A.20)). The unimodular condition on $\hat{\mathbf{n}}$ translates to the constraint

$$|\hat{\mathbf{n}}| = |\mathbf{z}|^2 = |z_1|^2 + |z_2|^2 = 1. \tag{14.3}$$

Some straightforward algebra is required to show that by (14.2) and (14.3) the following identity holds:

$$\frac{1}{4}|\partial_\mu\hat{\mathbf{n}}|^2 = (\partial_\mu\mathbf{z}^\dagger) \cdot (\partial_\mu\mathbf{z}) - [\mathcal{A}_\mu(\mathbf{z}^*, \mathbf{z})]^2 , \tag{14.4}$$

where $\mathcal{A}_\mu$ is the bilinear form:

$$\mathcal{A}_\mu \equiv -\frac{i}{2}\left[\mathbf{z}^\dagger\partial_\mu\mathbf{z} - (\partial_\mu\mathbf{z}^\dagger)\mathbf{z}\right] . \tag{14.5}$$

The measure of the NLSM partition function is

$$
\begin{aligned}
\mathcal{D}\hat{n} &= \prod_{\mathbf{x},m=1,2} \frac{d\operatorname{Re} z_m(\mathbf{x})\, d\operatorname{Im} z_m(\mathbf{x})}{\pi} \delta\left(|\mathbf{z}|^2 - 1\right) \\
&\equiv \mathcal{D}^2 \mathbf{z}\, \delta\left(|\mathbf{z}|^2 - 1\right).
\end{aligned}
\tag{14.6}
$$

The CP^1 representation is thus

$$
\begin{aligned}
Z_{NLSM} &= \int \mathcal{D}^2 \mathbf{z}\, \delta\left(|\mathbf{z}|^2 - 1\right) \\
&\quad \times \exp\left(-\frac{2\Lambda^{d-2}}{f} \int d^d x \sum_\mu |[\partial_\mu - i\mathcal{A}_\mu(\mathbf{z}^*, \mathbf{z})]\,\mathbf{z}|^2\right).
\end{aligned}
\tag{14.7}
$$

The constraints in the measure and the interactions in the action are the sources of difficulty in solving for the correlations of (14.7). We can replace the δ-function constraints by introducing a real constraint field $\lambda(\mathbf{x})$ and the identity

$$
\delta\left(|\mathbf{z}|^2 - 1\right) = \int \mathcal{D}\lambda \exp\left[i \int d^d x\, \lambda\left(|\mathbf{z}|^2 - 1\right)\right].
\tag{14.8}
$$

We can decouple the four-field interactions $\mathcal{A}_\mu^2$ by introducing auxiliary gauge fields A_μ and using the *Hubbard–Stratonovich* identity:

$$
\exp\left(c\mathcal{A}_\mu^2\right) = \sqrt{\frac{c}{\pi}} \int_{-\infty}^{\infty} dA_\mu\, \exp\left(-cA_\mu^2 - 2cA_\mu\mathcal{A}_\mu\right).
\tag{14.9}
$$

By (14.7), (14.8) and (14.9), the partition function is given (up to overall normalization) by

$$
\begin{aligned}
Z_{CP^1} &= \int \mathcal{D}^2 \mathbf{z}\mathcal{D}A\mathcal{D}\lambda \\
&\quad \exp\left\{-\int d^d x\left[\frac{2\Lambda^{d-2}}{f} \sum_\mu |(\partial_\mu - iA_\mu)\,\mathbf{z}|^2 - i\lambda(|\mathbf{z}|^2 - 1)\right]\right\}.
\end{aligned}
\tag{14.10}
$$

The integrand is Gaussian in the z fields. The price we pay is the additional gauge fields $\mathbf{A} = \{A_\mu\}$ and constraint field λ. Their spatial fluctuations introduce interactions between the $\mathbf{z}$ variables. The next section will describe the mean field approximation where these fluctuations are ignored at the saddle point.

14.2 CP^{N-1} Models at Large N

We first generalize the CP^1 model to the CP^{N-1} model by extending the number of complex fields from $2 \to N$:

$$\mathbf{z} = (z_1, z_2, \ldots, z_N), \qquad N \geq 2. \tag{14.11}$$

The constraint (14.3) is generalized to

$$|\mathbf{z}|^2 = \sum_{m=1}^{N} |z_m|^2 = \frac{N}{2}. \tag{14.12}$$

The N dependence of (14.12) ensures a nontrivial large N theory. Thus, (14.10) is generalized to

$$Z_{CP^{N-1}} = \int \mathcal{D}^2 \mathbf{z}\, \mathcal{D}\mathbf{A}\, \mathcal{D}\lambda \exp\left[-\frac{2\Lambda^{d-2}}{f} \int d^d x \sum_\mu |(\partial_\mu - i\mathbf{A}_\mu)\mathbf{z}|^2 \right.$$
$$\left. -i\lambda\left(|\mathbf{z}|^2 - \frac{N}{2}\right)\right]. \tag{14.13}$$

Formally, we can integrate out the $\mathbf{z}$ fields and obtain

$$Z_{CP^{N-1}} = \int \mathcal{D}\mathbf{A}\, \mathcal{D}\lambda \,\exp\left(-N\mathcal{S}[\mathbf{A}, \lambda]\right),$$
$$\mathcal{S}[\mathbf{A}, \lambda] = \text{Tr}\ln\left[-(\partial_\mu - iA_\mu)^2\right] - \frac{i\Lambda^{d-2}}{f}\int d^d x\, \lambda. \tag{14.14}$$

Note that N multiplies an N-independent action $\mathcal{S}$. As explained in Appendix E, this justifies, at large N, a saddle point approximation for (14.14):

$$Z \approx \exp\left(-N\mathcal{S}[\mathbf{A}_0, \lambda_0]\right) Z', \tag{14.15}$$

where the saddle point equations are

$$\left.\frac{\delta\mathcal{S}}{\delta\bar{\mathbf{A}}}\right|_{\bar{\mathbf{A}}_0,\lambda_0} = \left.\frac{\delta\mathcal{S}}{\delta\lambda}\right|_{\bar{\mathbf{A}}_0,\lambda_0} = 0. \tag{14.16}$$

Assuming no broken gauge or translational symmetry in the saddle point action $\mathcal{S}[\bar{\mathbf{A}}_0, \lambda_0]$, we take

$$\mathbf{A}_0 = 0,$$
$$\lambda_0 = -i\bar{\lambda}. \tag{14.17}$$

The saddle point action is the free energy of a noninteracting field z:

$$\mathcal{S}(0, \bar{\lambda}) = \sum_{\mathbf{k}} \ln\left[\mathbf{k}^2 + \bar{\lambda}\right] - \frac{N\Lambda^{d-2}}{f}\bar{\lambda}. \tag{14.18}$$

The saddle point equation for $\bar{\lambda}$ is

$$\mathcal{N}^{-1} \sum_{\mathbf{k}} \frac{1}{|\mathbf{k}|^2 + \bar{\lambda}} - \frac{1}{f} = 0, \qquad (14.19)$$

which yields

$$\bar{\lambda} = \begin{cases} f^2/4 & d = 1 \\ \frac{1}{2}\Lambda^2 \exp\left(-\frac{4\pi}{f}\right) & d = 2 \\ \frac{1}{2}\Lambda^2 \left(\frac{2}{\pi} - \frac{4\pi}{f}\right)^2 & d = 3 \end{cases}. \qquad (14.20)$$

We generalize the $O(3)$ correlation function (of $N = 2$) to $N > 2$ by defining

$$\begin{aligned} S^{+-}(\mathbf{x}) &\equiv \langle [\hat{n}^x(0) + i\hat{n}^y(0)] [\hat{n}^x(\mathbf{x}) - i\hat{n}^y(\mathbf{x})] \rangle \\ &\rightarrow \langle z_m^*(0) z_{m'}(0) z_{m'}^*(\mathbf{x}) z_m(\mathbf{x}) \rangle \\ &\equiv S^{m \neq m'}(\mathbf{x}). \end{aligned} \qquad (14.21)$$

Since the flavors are decoupled in the saddle point action (14.18), we obtain

$$\begin{aligned} S^{m \neq m'}(\mathbf{x}) &= |G(\mathbf{x})|^2 , \\ G(\mathbf{x}) &= Z^{-1} \int \mathcal{D}^2 \mathbf{z} \; \mathbf{z}^\dagger(0) \cdot \mathbf{z}(\mathbf{x}) \\ &\qquad \times \exp\left(-\frac{2\Lambda^{d-2}}{f} \sum_{\mathbf{k}} (|\mathbf{k}|^2 + \bar{\lambda}) \, z_{\mathbf{k}}^* z_{\mathbf{k}} - \frac{N\mathcal{N}\Lambda^{d-2}}{f} \bar{\lambda} \right) \\ &= \mathcal{N}^{-1} \sum_{\mathbf{k}} \frac{e^{-i\mathbf{k}\cdot\mathbf{x}}}{|\mathbf{k}|^2 + \bar{\lambda}}. \end{aligned} \qquad (14.22)$$

The Fourier sum can be evaluated asymptotically for large distances following (13.48),

$$G(\mathbf{x}) \sim \text{const} \; |\mathbf{x}|^{-(d-1)/2} \exp\left(-|\mathbf{x}|\sqrt{\bar{\lambda}}\right). \qquad (14.23)$$

Taking the square of $G(\mathbf{x})$, we obtain the asymptotic correlation function

$$\lim_{|\mathbf{x}| \gg \xi} S^{m \neq m'}(\mathbf{x}) \sim \text{const} \; |\mathbf{x}|^{-(d-1)} \exp\left(-|\mathbf{x}|/\xi\right), \qquad (14.24)$$

where

$$\xi = \frac{1}{2\sqrt{\bar{\lambda}}}. \qquad (14.25)$$

By (14.20) we find that

$$
\xi = \begin{cases} \frac{\Lambda^{-1}}{f} & d = 1 \\ \frac{1}{2}\Lambda^{-1} \exp\left(+\frac{2\pi}{f}\right) & d = 2 \\ \frac{1}{2}\Lambda^{-1}\left(\frac{2}{\pi} - \frac{4\pi}{f}\right)^{-1} & d = 3 \end{cases}. \tag{14.26}
$$

These results (applied to the $N = 2$ case) reproduce the correlation lengths of the $O(3)$ NLSM given by (13.54) and (13.55). That is to say, the large N approximation coincides with the weak coupling renormalization group results up to one loop order.

It is apparent that the large N theory of the CP^{N-1} is not at all bad for describing the prominent long-wavelength features of the Heisenberg model. This is reassuring since large N approximations are often applied in condensed matter theory and particle physics to describe physically interesting systems with $N = 2$.[1] We must note, however, that the large N approximation does not recover the preexponential dependence of $\xi(f)$ and $S^{+-}(\mathbf{x})$ found in more detailed two loop order renormalization group results. These differences are presumably due to Z', which includes the higher-order terms in the $1/N$ expansion.

14.3 Exercises

1. Prove that for the CP^1 model

$$
(\partial_\mu A_\nu - \partial_\nu A_\mu) = \frac{1}{2}\partial_\mu \hat{\mathbf{n}} \times \partial_\nu \hat{\mathbf{n}} \cdot \hat{\mathbf{n}}, \tag{14.27}
$$

where $A(\mathbf{z})$ is the CP^1 gauge field defined by (14.5).

2. Compare (13.61) and (14.27), and show that the definitions (13.58) and (14.5) of the gauge fields are equivalent up to a total derivative term. Use this comparison to show that the *charge* of the z spinors is half the charge of the spin wave modes ψ.

Bibliography

The $1/N$ expansion of CP^{N-1} models was introduced by

- A. d'Adda, M. Luscher, and P. di Vecchia, Nucl. Phys. B **146**, 63 (1978); **152**, 125 (1979).

Pedagogical texts on large N expansion of field theories are

---

[1]See Chapters 16 and 18 for large N theories of lattice Heisenberg models.

- S. Coleman, *Aspects of Symmetry* (Cambridge University Press, 1990), Chapter 8;
- A.M. Polyakov, *Gauge Fields and Strings* (Harwood, 1987), Chapter 8.

15

Quantum Antiferromagnets: Continuum Results

In Chapter 12, we have used Haldane's mapping to relate the long-wavelength correlations of the quantum Heisenberg antiferromagnet (QHA) in d dimensions to those of the Nonlinear Sigma Model (NLSM) in $d+1$ dimensions with additional Berry phases. In Chapters 13 and 14, the correlation length ξ for the NLSM was found for $d=1,2,3$ using the continuum approximation.

In principle, Haldane's mapping holds when the continuum approximation is justified, i.e.,

$$\xi_{NLSM} \gg \Lambda^{-1}, \quad \xi_{QHA} \gg a, \tag{15.1}$$

where Λ and a are the high-momentum cutoff and lattice constant, respectively.

Before we can use the NLSM results of Chapters 13 and 14, we must understand the effects of the Berry phase term $e^{i\Upsilon}$ on the right-hand side of (12.37).

15.1 One Dimension, the Θ Term

We examine the Berry phase Υ defined in (12.28) for the one-dimensional chain. For slowly varying $\hat{n}(x)$, this term is expanded:

$$
\begin{aligned}
\Upsilon^{d=1} &= -S \sum_i \omega\left[\hat{n}(x_{2i})\right] - \omega\left[\hat{n}(x_{2i-1})\right] \\
&= \frac{S}{2} \int \frac{dx}{a} \frac{\delta\omega}{\delta\hat{n}} \cdot \partial_x\hat{n}\, a \\
&= 2\pi S\Theta\left[\hat{n}(x,\tau)\right].
\end{aligned}
\tag{15.2}
$$

We use (10.34) to derive the functional form of Θ as

$$\Theta = \frac{1}{4\pi} \int d\tau \int dx\, (\hat{n} \times \partial_\tau\hat{n} \cdot \partial_x\hat{n}). \tag{15.3}$$

Θ is the topological "winding number" or "Pontryagin index" of the mapping

$$\hat{n} : \left\{[-\frac{L}{2}, \frac{L}{2}), [0, \beta)\right\} \rightarrow S^2, \tag{15.4}$$

where S^2 is the surface of the unit sphere. $\Theta[\hat{n}]$ is the area of the range of $\hat{n}$ divided by 4π. If the domain has periodic boundary conditions, $\hat{n}$ must cover the sphere an integer number of times, and therefore Θ must be an integer. This integer is topologically stable, i.e., it cannot be changed by any continuous deformation of $\hat{n}$. Changing Θ from 1 to 0 is analogous to unwrapping a sphere, which is impossible unless one makes a hole or a tear in the wrapping paper.

The Θ term also appears in the path integral representation of the Green's function, where $\tau \to it$ (see the Exercises). As a topological invariant its variation is zero, and it cannot affect the classical equations of motion. The phase factor $e^{i2\pi S\Theta}$ is unity for all integer spins, but it can be positive or negative for half-odd integer spins $S = \frac{1}{2}, \frac{3}{2}, \dots$. This causes interference effects in the path integral, which dramatically alter the ground state correlations and excitations from that of the pure NLSM.

The continuum theory in the presence of alternating Berry phases is hard to solve. However, we know from the Lieb, Schultz, and Mattis theorem (see Section 5.2), that the lowest excitation energy of the half-odd integer spin chain vanishes as the inverse number of sites. From Bethe's solution of the integrable $S = \frac{1}{2}$ chain, the spectrum is known to be gapless, and the ground state correlations decay asymptotically as

$$\langle \mathbf{S}_i \cdot \mathbf{S}_j \rangle \sim (-1)^{(i-j)} \frac{(\ln|i-j|)^{\frac{1}{2}}}{|i-j|}. \tag{15.5}$$

Thus we can conclude that the Θ term destroys Haldane's gap and changes the asymptotic correlations.[1]

Now we present a heuristic argument for the relation between a topological interference term and the ground state degeneracy. Consider a spinless particle moving on a circle pierced by a magnetic flux tube of flux ϕ as depicted in Fig. 15.1. The Hamiltonian is simply

$$\mathcal{H} = \frac{\hbar^2}{2m} \left(-i\partial_x + \frac{\phi}{L\phi_0} \right)^2, \tag{15.6}$$

where $\phi_0 = hc/me$ is the flux quantum. The spectrum is labelled by integers n (see Fig. 15.1),

$$E(n) = \frac{\hbar^2}{2mL^2} \left(n + \frac{\phi}{\phi_0} \right)^2, \quad \pm n = 0, 1, 2 \dots. \tag{15.7}$$

For integer ϕ, the ground state is

$$|n_0\rangle = |-\phi/\phi_0\rangle, \tag{15.8}$$

---

[1] See Affleck, and Shankar and Read.

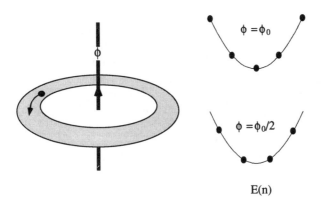

FIGURE 15.1. Analogy of the Θ term: a particle on a ring pierced by a magnetic flux tube. Half-odd integer flux quanta yield ground state degeneracy.

and it is nondegenerate. On the other hand, for half-odd integer ϕ/ϕ_0, the ground states are the two degenerate states

$$| - \phi/\phi_0 - \frac{1}{2}\rangle \, , \quad | - \phi/\phi_0 + \frac{1}{2}\rangle. \tag{15.9}$$

The analogy between the particle on a circle and the two-dimensional NLSM with a Θ term is seen in the path integral representation of (15.6),

$$Z \; = \int \mathcal{D}x \, e^{i\Upsilon[x(\tau)]} \, \exp\left(-\frac{m}{\hbar} \int_0^\beta d\tau \, \frac{\dot{x}^2}{2} \right), \tag{15.10}$$

where Υ is the pure gauge coupling between the particle and the flux tube (i.e., an Aharonov–Bohm phase)

$$
\begin{aligned}
\Upsilon \; &= \; \frac{2\pi\phi}{L\phi_0} \int_0^\beta d\tau \, \dot{x} \\
&= \; 2\pi n \frac{\phi}{\phi_0}, \quad \pm n = 0, 1, 2\ldots.
\end{aligned}
\tag{15.11}
$$

n is the winding number of $x(\tau)$, i.e., the number of times the particle has circulated around the ring. It is the one-dimensional analogue of the two-dimensional Pontryagin index Θ of (15.3). Thus for half-odd integer flux quantums ϕ and half-odd integer spins S, the two systems have destructive interference between even and odd topological sectors. In both systems this interference is responsible for their ground state degeneracies.

15.2 One Dimension, Integer Spins

By (15.2), we can ignore the Berry phase factor $e^{i\Upsilon}$ for one-dimensional quantum Heisenberg antiferromagnets (QHA) of integer spins. By Haldane's mapping, the continuum approximation to the QHA is the $d=2$ NLSM which by Section 13.1 is the continuum limit of the classical Heisenberg model. By Mermin and Wagner's Theorem 6.2, the latter has no long-range order for all finite temperatures, which translates to no long-range order for the QHA ground state for all $S < \infty$.

The correlation length $\xi_{NLSM}(f)$ was evaluated in (13.55). The same result was also obtained by the large N approximation (14.26). Thus,

$$
\begin{aligned}
\xi_{QHA}(d=1,S) &= \xi_{NLSM}(d=2,f) \\
&\propto a \exp\left(\frac{2\pi}{f}\right), \quad S = 0,1,\dots. \quad (15.12)
\end{aligned}
$$

By Table 12.1, the coupling constant f for the nearest neighbor model is

$$
f = 2/S, \quad (15.13)
$$

which yields

$$
\begin{aligned}
\xi_{QHA}(S,T=0) &\approx a \exp(\pi S), \\
\Delta &\approx \frac{c}{a} \exp(-\pi S) \quad (15.14)
\end{aligned}
$$

where Δ is Haldane's gap given by (12.39). Equation (15.14) demonstrates that correlation length and Haldane's gap are *nonperturbative* in the semiclassical parameter, i.e., they cannot be obtained as a spin wave expansion in powers of $1/S$. The formation of a disordered (spin liquid) ground state with a gap to all excitations is a purely quantum phenomenon with no classical analogue. It is similar in that respect to tunneling of a single particle under an energy barrier.[2]

We recall that ground states of the AKLT Heisenberg models in one dimension (see Chapter 8, (8.26)) were found to have exponentially decaying correlations (8.30). Also, within the single-mode approximation (see Section 9.3), we have estimated their gap in (9.21). Thus, the AKLT model of $S = 1$ with its valence bonds ground states is qualitatively similar to other integer spin Heisenberg antiferromagnets, which map onto the NLSM. Since the AKLT models have S dependent interactions, they scale differently from the standard Heisenberg models at large S.

---

[2]Tunneling rates go as $\exp\left[-\frac{1}{\hbar}(\cdot)\right]$, where $\hbar$ is the semiclassical parameter.

15.3   Two Dimensions

At zero temperature, the QHA on the square lattice maps onto the three-dimensional NLSM, which, according to (13.46), has a disordered phase at

$$f_c \geq 2\pi^2. \tag{15.15}$$

For the nearest neighbor (nn) model, Neves and Perez proved that the ground state is ordered for all $S \geq 1$. In addition, series expansions and numerical simulations indicate an ordered ground state for $S = \frac{1}{2}$.

Unfortunately, the precise determination of $f(S, \Lambda)$ for extended Heisenberg models depends on the details of the short-wavelength regularization procedure (i.e., is "nonuniversal"). We recall that the continuum approximation itself depends on inequalities (12.10) and (15.16). Since near the critical temperature $T_c = f_c J S^2$ even the nearest neighbor correlations are substantially degraded, the use of the NLSM field theory is questionable in this regime.

The first corrections to the continuum approximation involve point singularities in the Néel field. These "hedgehogs" are expected to be thermally excited in the disordered phase. Haldane has shown[3] that hedgehogs introduce nontrivial Berry phases $\Upsilon(2S \bmod 4)$, which may give rise to ground state degeneracy, for $S = \frac{1}{2}, 1$, but not, e.g., $S = 2$. Large N approaches by Read and Sachdev agree with Haldane's scenario and predict spin Peierls ordering due to the hedgehog Berry phases.[4]

By symmetry of the NLSM under rotations in $d+1$ dimensions, for a d-dimensional momentum cutoff Λ, the shortest timescale in the path integral is $2\pi/(c\Lambda)$. Thus, in the regime

$$T << c\Lambda, \tag{15.16}$$

the correlation length can be calculated by the renormalization group equations for the $d=3$ NLSM on a finite *slab of width* $c\beta$. This was carried out by Chakaravarty, Halperin, and Nelson (CHN) up to two-loop order in the β function. CHN found that in the weak coupling regime, f flows to the Néel fixed point $f_c = 0$, and the ground state has long-range order. At finite temperatures, they calculated the correlation length to be

$$\xi_{QHA}^{d=2} = 0.9(c/T) \exp\left(\frac{2\pi\rho_s}{T}\right). \tag{15.17}$$

We can compare $\xi_{QHA}^{d=2}$ to the classical result for the two-dimensional Heisenberg model given by (13.5) and (13.55). The essential effect of the quantum fluctuations is to renormalize the stiffness constant by a constant Z_ρ:

$$\rho_s = Z_\rho \rho_s^{cl} . \tag{15.18}$$

---

[3]See bibliography of Chapter 12.
[4]See bibliography of Chapter 18.

The correlation length (15.17) is called a *renormalized classical* correlation. Z_ρ for $S = 1/2$ was calculated by series expansions[5] to be 0.183.

Many of the predictions of the NLSM for the $S = \frac{1}{2}$ QHA have been confirmed both numerically and experimentally in quasi-two-dimensional systems of La_2CuO_4. The experimental and theoretical aspects of possible quantum disordered ground states (quantum spin liquids) in two dimensions are still unsettled.

Bibliography

A proof that the NLSM with a Θ term is gapless for half-odd integer S was given by

- R. Shankar and N. Read, Nucl. Phys. B **336**, 457 (1990).

For a review on quantum spin chains see

- I. Affleck, J. Phys. Condensed Matter **1**, 3047 (1989).

Proof of long-range order for the ground state of the Heisenberg antiferromagnet on the square lattice for $S \geq 1$ was given by

- E.J. Neves and J.F. Perez, Phys. Lett. A **114**, 331 (1986).

The finite temperature correlations in two dimensions have been evaluated using the renormalization group by

- S. Chakravarty, B.I. Halperin, and D. Nelson, Phys. Rev. B **39**, 2344 (1989).

Two extensive reviews on the two-dimensional Heisenberg antiferromagnet of spin half and layered cuprates are

- S. Chakravarty, *"High Temperature Superconductivity"*, edited by K.S. Bedell, D. Coffey, D.E. Meltzer, D. Pines, and J.R. Schrieffer (Addison-Wesley, 1990);
- E. Manousakis, Rev. Mod. Phys. **63**, 1 (1991).

[5]See R.R.P. Singh, Phys. Rev. B **39**, 9760 (1989).

16

SU(N) Heisenberg Models

The use of large N approximations to treat strongly interacting quantum systems been very extensive in the last decade. The approach originated in elementary particles theory, but has found many applications in condensed matter physics (see bibliography). Initially, the large N expansion was developed for the Kondo and Anderson models of magnetic impurities in metals. Soon thereafter it was extended to the Kondo and Anderson lattice models for mixed valence and heavy fermions phenomena in rare earth compounds. The approach was also applied to treat the strong Coulomb interactions in high T_c cuprate superconductors.

In the present chapter and in Chapters 17 and 18, we shall formulate and apply the large N approach to the quantum Heisenberg model. This method provides an additional avenue to the static and dynamical correlations of quantum magnets. The mean field theories derived below can describe both ordered and disordered phases, at zero and at finite temperatures, and they complement the semiclassical approaches of Chapters 11–15.

Generally speaking, the parameter N labels an internal SU(N) symmetry at each lattice site (i.e., the number of "flavors" a Schwinger boson or a constrained fermion can have). In most cases, the large N approximation has been applied to treat spin Hamiltonians, where the symmetry is SU(2), and N is therefore not a truly large parameter. Nevertheless, the $1/N$ expansion provides an easy method for obtaining simple mean field theories. These have been found to be either surprisingly successful or completely wrong, depending on the system. For example: we shall see in Chapter 18 that the Schwinger boson mean field theory in one dimension works well for the ferromagnet and for the antiferromagnet of integer spin but fails for the half-odd integer spin antiferromagnet.

The large N approach handles strong local interactions in terms of constraints. It is not a perturbative expansion in the size of the interactions but rather a saddle point expansion which usually preserves the spin symmetry of the Hamiltonian. At the mean field level, the constraints are enforced only on average. Their effects are systematically reintroduced by the higher-order corrections in $1/N$.

The $1/N$ corrections to mean field theory can be calculated by Feynman diagrams, which describe interactions between the free quasiparticles of the mean field theory. In Section 17.3, we use the unique structure of the $1/N$ expansion to derive certain sum rules to all orders.

It turns out that different large N generalizations are suitable for different Heisenberg models, depending on the sign of couplings, spin size, and lattice. By "suitable" we mean that the corresponding mean field theories produce the correct qualitative features of the ground states. Below, we introduce three large N generalizations of the Heisenberg model. We begin by the Schwinger boson representation of the ferromagnet.

16.1   Ferromagnet, Schwinger Bosons

Following Section 7.2, we introduce two Schwinger bosons per site a_i, b_i and enforce the local constraints on their Fock space,

$$a_i^\dagger a_i + b_i^\dagger b_i = 2S. \tag{16.1}$$

A bond operator is defined as

$$\mathcal{F}_{ij} \equiv a_i^\dagger a_j + b_i^\dagger b_j. \tag{16.2}$$

The ferromagnetic Heisenberg model of spin S is written as

$$
\begin{aligned}
\mathcal{H} &= -J \sum_{\langle ij \rangle} \mathbf{S}_i \cdot \mathbf{S}_j \\
&= -\frac{J}{2} \sum_{\langle ij \rangle} \left[\mathcal{F}_{ij}^\dagger \mathcal{F}_{ij} - 2S(S+1) \right] \\
&= -\frac{J}{2} \sum_{\langle ij \rangle} \left(: \mathcal{F}_{ij}^\dagger \mathcal{F}_{ij} : -2S^2 \right).
\end{aligned}
\tag{16.3}
$$

The sum $\sum_{\langle ij \rangle}$ is over *bonds* with endpoints (i,j), and : : denotes normal ordering (see (C.18)). Equation (16.3) is generalized to SU(N) by increasing the number of Schwinger boson flavors from 2 to N:

$$(a_i, b_i) \rightarrow (a_{i1}, a_{i2}, \ldots, a_{iN}). \tag{16.4}$$

The bond operator is generalized to

$$\mathcal{F}_{ij} \rightarrow \sum_{m=1}^{N} a_{im}^\dagger a_{jm}, \tag{16.5}$$

and the constraints are

$$\sum_{m=1}^{N} a_{im}^\dagger a_{im} = NS = 0, 1, 2, \ldots. \tag{16.6}$$

The SU(N) *ferromagnetic boson* (FM-B) Heisenberg model is

$$
\begin{aligned}
\mathcal{H}^{FM-B}(N) &= -\frac{J}{N} \sum_{\langle ij \rangle} \left(: \mathcal{F}_{ij}^\dagger \mathcal{F}_{ij} : -NS^2 \right) \\
&= -\frac{J}{N} \sum_{ij} \left(\sum_{mm'} S_i^{mm'} S_j^{m'm} - NS^2 \right),
\end{aligned}
\tag{16.7}
$$

where the second line is given by rearranging the operators and defining the SU(N) generators as

$$
S^{mm'} = a_m^\dagger a_{m'}.
\tag{16.8}
$$

$S^{mm'}$ obey the SU(N) algebra

$$
\left[S^{mm'}, S^{\mu\mu'} \right] = \delta_{m'\mu} S^{m\mu'} - \delta_{m\mu'} S^{\mu m'}.
\tag{16.9}
$$

A particular "spin size" S is defined by the constraint (16.6). For SU(2) we recognize the usual spin operators as $S^{12} = S^+$, $S^{21} = S^-$, and $S^{11} = S^z + S$.

It can be verified that $\mathcal{H}^{FM-B}$ is invariant under uniform $SU(N)$ transformations on the spins. We have intentionally written the Hamiltonian as a *negative* quadratic form $-\mathcal{F}_{ij}^\dagger \mathcal{F}_{ij}$. As we shall see, this allows a natural starting point for decoupling the interactions in the mean field theory.

16.2 Antiferromagnet, Schwinger Bosons

We consider the case of nearest neighbor antiferromagnetic interaction $J > 0$, on a bipartite lattice with sublattices A, B. A bond $\langle ij \rangle$ is defined such that $i \in A$ and $j \in B$. The antiferromagnetic bond operator is defined as

$$
\vec{\mathcal{A}}_{ij} = a_i b_j - b_i a_j.
\tag{16.10}
$$

The arrow $\rightarrow$ denotes the antisymmetry with respect to interchange of $i \rightarrow j$. We define a spin rotation by π about the y axis on sublattice B which sends

$$
a_j \rightarrow -b_j , \qquad b_j \rightarrow a_j .
\tag{16.11}
$$

This is a canonical transformation which preserves the constraint (16.1). The antiferromagnetic bond operator transforms into a symmetric operator:

$$
\vec{\mathcal{A}}_{ij} \rightarrow \mathcal{A}_{ij} = a_i a_j + b_i b_j.
\tag{16.12}
$$

The SU(2) Heisenberg model is written in the form

$$
\mathcal{H} = J \sum_{\langle ij \rangle} \mathbf{S}_i \cdot \mathbf{S}_j
$$

$$= -\frac{J}{2} \sum_{\langle ij \rangle} \left(\mathcal{A}_{ij}^\dagger \mathcal{A}_{ij} - 2S^2 \right). \tag{16.13}$$

As in the ferromagnetic case, H can be generalized to $N > 2$ models by adding Schwinger boson flavors. The constraint is generalized to (16.6), and the bond operator is generalized to

$$\mathcal{A}_{ij} \to \sum_{m=1}^{N} a_{im} a_{jm}. \tag{16.14}$$

The SU(N) antiferromagnetic bosons (AFM-B) Heisenberg model is

$$\mathcal{H}^{AFM-B}(N) = -\frac{J}{N} \sum_{\langle ij \rangle} \left(\mathcal{A}_{ij}^\dagger \mathcal{A}_{ij} - NS^2 \right)$$

$$= -\frac{J}{N} \sum_{\langle ij \rangle} \left(\sum_{mm'} S_i^{mm'} \tilde{S}_j^{m'm} - NS^2 \right), \tag{16.15}$$

where

$$\tilde{S}_j^{mm'} = a_{jm'}^\dagger a_{jm} \tag{16.16}$$

are the generators of the *conjugate representation* on sublattice B. One should note that $\mathcal{H}^{AFM-B}$ of (16.15) is not invariant under uniform $SU(N)$ transformations U but only under staggered conjugate rotations U and $U^\dagger$ on sublattices A and B, respectively.

16.3   Antiferromagnet, Constrained Fermions

In Chapter 2, we first encountered quantum spins which are bilinear electron operators. Spin half operators can be represented by two states per site $|i, \uparrow\rangle, |i, \downarrow\rangle$, which are occupied by a single fermion, i.e.,

$$n_{i\uparrow} + n_{i\downarrow} = 1 \,\, (= 2S). \tag{16.17}$$

This representation is limited to $S = \frac{1}{2}$ due to the Pauli principle.[1] The spin operators are (see (2.10))

$$\begin{aligned} S_i^+ &= c_{i\uparrow}^\dagger c_{i\downarrow}, \\ S_i^- &= c_{i\downarrow}^\dagger c_{i\uparrow}, \\ S_i^z &= \frac{1}{2}(c_{i\uparrow}^\dagger c_{i\uparrow} - c_{i\downarrow}^\dagger c_{i\downarrow}). \end{aligned} \tag{16.18}$$

---

[1] Representations of $S > \frac{1}{2}$ can be constructed by adding orbitals and additional constraints.

The bond operator is defined as

$$\mathcal{D}_{ij} = c_{i\uparrow}^{\dagger} c_{j\uparrow} + c_{i\downarrow}^{\dagger} c_{j\downarrow}, \tag{16.19}$$

and the Heisenberg model is written as

$$
\begin{aligned}
\mathcal{H} &= J \sum_{\langle ij \rangle} \mathbf{S}_i \cdot \mathbf{S}_j \\
&= -\frac{J}{2} \sum_{\langle ij \rangle} : \mathcal{D}_{ij}^{\dagger} \mathcal{D}_{ij} : .
\end{aligned} \tag{16.20}
$$

It is possible to generalize representation (16.18) to SU(N) by introducing N flavors of fermions, c_{im}, $m = 1, \ldots, N$, such that

$$S^{mm'} = c_m^{\dagger} c_{m'}, \tag{16.21}$$

which obey the SU(N) algebra (16.9). The constraint on the fermions is generalized to

$$\sum_{m=1}^{N} c_{im}^{\dagger} c_{im} = NS, \tag{16.22}$$

where S is the generalized "spin size." The bond operator is defined by

$$\mathcal{D}_{ij} = \sum_m c_{im}^{\dagger} c_{jm}. \tag{16.23}$$

The SU(N) antiferromagnetic fermions (AFM-F) Heisenberg model is given by generalizing (16.20),

$$
\begin{aligned}
\mathcal{H}^{AFM-F}(N) &= \frac{J}{N} \sum_{\langle ij \rangle} \sum_{mm'} S_i^{mm'} S_j^{m'm} \\
&= -\frac{J}{N} \sum_{\langle ij \rangle} : \mathcal{D}_{ij}^{\dagger} \mathcal{D}_{ij} : .
\end{aligned} \tag{16.24}
$$

16.4 The Generating Functional

The generating Hamiltonian is defined by

$$\mathcal{H}[j] = \mathcal{H} - \sum_{imm'} j_{imm'}(\tau)\, a_{im}^{\dagger} a_{im'}, \tag{16.25}$$

where $\tau \in [0, \beta)$ is the imaginary time. The constraints (16.6) or (16.22) are enforced by the projector P_S, which commutes with $\mathcal{H}[j]$. The imaginary time generating functional (see (D.1)) is

$$
Z[j] \;=\; \mathrm{Tr}\, P_S T_\tau \left[\exp\left(-\int_0^\beta d\tau\, \mathcal{H}[j] \right) \right]
$$

$$
=\; \lim_{\epsilon \to 0} \mathrm{Tr}\, T_\tau \prod_{\tau_n = \epsilon}^{\beta} \left[P_S(\tau) \exp\left(-\epsilon \mathcal{H}[j(\tau_n)] \right) \right]. \tag{16.26}
$$

We use an integral representation of the constraint

$$
P_S(\tau) \;=\; \int \mathcal{D}\lambda \, \exp\left[-i\epsilon \sum_{im} \lambda_i(\tau)\, (a_{im}^\dagger a_{im} - S) \right], \tag{16.27}
$$

where the measure of the constraint field is

$$
\int \mathcal{D}\lambda \;=\; \lim_{\epsilon \to 0} \prod_{i\tau} \epsilon \int_{-\pi/\epsilon}^{\pi/\epsilon} d\lambda_{i\tau}. \tag{16.28}
$$

The exponential of (16.27) and $\mathcal{H}[j]$ can be combined in the exponent since they commute.

Following Appendix D, we construct a coherent states path integral for the generating functional which has unified notations for the Schwinger bosons and constrained fermion Hamiltonians:

$$
Z[j] \;=\; \int_{-\infty}^{\infty} \mathcal{D}\lambda \int \mathcal{D}^2 \mathbf{z} \exp\left\{ -\int_0^\beta d\tau \left[\sum_{im} z_{im}^* \partial_\tau z_{im} + H[j] \right. \right.
$$

$$
\left. \left. + i \sum_{im} \lambda_i(\tau)(z_{im}^* z_{im} - S) \right] \right\}, \tag{16.29}
$$

where $\mathbf{z}$ are complex variables for the bosons or Grassmann variables for the fermions. The Hamiltonian function is

$$
H[j] \;=\; -\frac{J}{N} \sum_{\langle ij \rangle} \mathcal{Z}_{ij}^* \mathcal{Z}_{ij} \;-\; \sum_{imm'} j_{imm'}(\tau) z_{im}^* z_{im'}, \tag{16.30}
$$

where

$$
\mathcal{Z} = \begin{cases} \sum_m z_{im}^* z_{jm} & \text{FM-B} \\ \sum_m z_{im} z_{jm} & \text{AFM-B} \\ \sum_m z_{im}^* z_{jm} & \text{AFM-F} \end{cases}, \tag{16.31}
$$

where the FM-B, AFM-B, and AFM-F Hamiltonians were defined in (16.7), (16.15), and (16.24), respectively.

16.5   The Hubbard–Stratonovich Transformation

Biquadratic (four-variable) terms in the Lagrangian cannot be readily integrated in the path integral. These terms can be decoupled into bilinear terms using the *Hubbard–Stratonovich identity*

$$
\exp\left[\frac{J}{N}\mathcal{Z}^*\mathcal{Z}\epsilon\right] = \int_{-\infty}^{\infty} d^2Q
$$
$$
\times \exp\left[-\left(\mathcal{Z}^*Q + \mathcal{Z}Q^* + N\frac{|Q|^2}{J}\right)\epsilon\right],
$$
(16.32)

where

$$
d^2Q \equiv \frac{\epsilon N}{\pi J}\operatorname{Re}Q\, d\operatorname{Im}Q \ . \tag{16.33}
$$

A complex variable $Q_{ij}(\tau)$ is introduced for each bond $\langle ij\rangle$ at each timestep, and the integration measure is defined as

$$
\mathcal{D}^2Q \equiv \prod_{\langle ij\rangle\tau} d^2Q_{ij}(\tau). \tag{16.34}
$$

Using (16.27), (16.29), and (16.32), we obtain

$$
Z[j] = \int \mathcal{D}^2Q\, \mathcal{D}\lambda\, \mathcal{D}^2z
$$
$$
\times \exp\left[-\int_0^\beta d\tau\left(L[j] + N\frac{|Q_{ij}|^2}{J} - iNS\sum_i \lambda_i\right)\right],
$$
(16.35)

where the Lagrangian is quadratic in the z variables,

$$
L[\mathbf{z}^*, \mathbf{z}, j] = \sum_{im} z_{im}^*\partial_\tau z_{im} + \sum_{\langle ij\rangle}\left(\mathcal{Z}_{ij}^\dagger Q_{ij} + \mathcal{Z}_{ij}Q_{ij}^*\right)
$$
$$
- \sum_{imm'}(j_{imm'} + i\lambda_i\delta_{mm'})z_{im}^* z_{im'}. \tag{16.36}
$$

For *normal* Lagrangians which contain only z^*z terms, e.g., the FM-B and AFM-F cases, the Green's function matrix $\hat{G}$ is defined by

$$
L = \mathbf{z}^*\hat{G}^{-1}[j]\mathbf{z} ,
$$
$$
\hat{G}^{-1} = \partial_\tau + \hat{\lambda} + \hat{Q} - \hat{j}, \tag{16.37}
$$

where ∂_τ is defined in (D.9) and $\hat{\lambda}, \hat{Q}, \hat{j}$ are read from (16.36).

For *anomalous* Lagrangians, such as the AFM-B, one has

$$\mathcal{L} = (\mathbf{z}_A^*, \mathbf{z}_B)\hat{G}^{-1}[j]\begin{pmatrix} \mathbf{z}_A \\ \mathbf{z}_B^* \end{pmatrix},$$

$$\hat{G}^{-1} = \begin{pmatrix} \partial_\tau + \hat{\lambda} - \hat{j} & \hat{Q} \\ \hat{Q}^\dagger & -\partial_\tau + \hat{\lambda} - \hat{j} \end{pmatrix}, \qquad (16.38)$$

where $\mathbf{z}_A, \mathbf{z}_B$ are the variables of sublattices A and B, respectively.

Following (C.13) and (D.11), we integrate the z variables and obtain a path integral of the auxiliary fields,

$$Z[j] = \int \mathcal{D}^2 Q \, \mathcal{D}\lambda \exp\left[-N\mathcal{S}[\lambda, Q, j]\right]$$

$$\mathcal{S} = -\frac{\eta}{N}\text{Tr}_{\tau im}\left(\ln \hat{G}[j]\right) + \int_0^\beta d\tau \left(\sum_{\langle ij \rangle} \frac{|Q_{ij}|^2}{J} - iS\sum_i \lambda_i\right).$$

$$(16.39)$$

This expression is the starting point for a steepest descents expansion controlled by N as the large parameter. The large N expansion will be reviewed in Chapters 17 and 18.

16.6   Correlation Functions

The connected two spins correlation function is given by

$$\tilde{R}^{mm'}(1,2) = \langle a_{m,1}^\dagger a_{m',1} a_{m',2}^\dagger a_{m,2} \rangle$$

$$= Z^{-1}\frac{\delta^2 \ln Z}{\delta j_{mm',1} \delta j_{m'm,2}}\bigg|_{j=0}$$

$$= Z^{-1}\int \mathcal{D}^2 Q \, \mathcal{D}\lambda \left[-N\frac{\delta^2 \mathcal{S}}{\delta j_{mm',1} \delta j_{m'm,2}}\right.$$

$$\left. +N^2 \frac{\delta \mathcal{S}}{\delta j_{mm',1}} \frac{\delta \mathcal{S}}{\delta j_{m'm,2}} \exp\left(-N\mathcal{S}[\lambda, Q]\right)\right],$$

$$(16.40)$$

where $1, 2$ are points in space-time, and all functions in the integrand are evaluated at $j = 0$. By the symmetry of $\mathcal{S}$ with respect to flavor indices, the second term is proportional to $\delta_{mm'}$, and $\tilde{R}^{m \neq m'}$ is independent on m, m'. For SU(2), the FM-B and AFM-F correlations are related to the usual spin correlations by

$$\tilde{R}_{N=2}^{m \neq m'}(1,2) = \langle S_{i_1}^+(\tau_1) S_{i_2}^-(\tau_2) \rangle. \qquad (16.41)$$

For the AFM-B case, the sublattice rotation (16.11) implies that

$$\tilde{R}^{m\neq m'}_{N=2}(1,2) = \begin{cases} \langle S^+_{i_1}(\tau_1) S^-_{i_2}(\tau_2) \rangle & i_1, i_2 \in A \\ -\langle S^+_{i_1}(\tau_1) S^+_{i_2}(\tau_2) \rangle & i_1 \in A,\, i_2 \in B \end{cases}. \tag{16.42}$$

Bibliography

Large N methods originated in field theory; see Chapter 14 and its bibliography.

The first large N expansion for interacting electrons was applied to the Coqblin–Schrieffer model (a large N generalization of the Kondo model) by

- N. Read and D. Newns, J. Phys. C **16**, 3273 (1983).

Coleman used the large N expansion for the slave boson representation of the Anderson model:

- P. Coleman, Phys. Rev. B **35**, 5072 (1987).

The use of $1/N$ expansion to derive Fermi liquid parameters for heavy fermions is found in

- A. Auerbach and K. Levin, Phys. Rev. Lett. **57**, 877 (1986);
- A. Millis and P. Lee, Phys. Rev. B **35**, 3394 (1987).

For extensive reviews on applications of large N approximations for heavy fermions and cuprate superconductors, see

- A. Hewson, *The Kondo Problem To Heavy Fermions* (Cambridge, 1993);
- N.E. Bickers, Rev. Mod. Phys. **59**, 845 (1987);
- G. Kotliar, *Correlated Electron Systems*, edited by V.J. Emery (World Scientific, 1993);
- K. Levin, J.H. Kim, J.P. Lu, and Q. Si, Physica C **175**, 449 (1991).

17

The Large N Expansion

In Chapter 16, we derived the functional integral representations of the generating functional and correlation functions for three families of SU(N) Heisenberg models. Expressions (16.39) and (16.40) are exact, but the integration over Q and λ involves a complicated nongaussian weight $\exp(-N\mathcal{S})$. The path integral can be expanded by the method of steepest descents as explained in Appendix E, where N is the large control parameter. A large value of N suppresses contributions from large fluctuations about the saddle point. N plays a role similar to the spin size S in the semiclassical expansion.[1] In contrast, however, the large N mean field theory is by no means "classical": it includes the disordering effects of quantum fluctuations.

The saddle point configurations are denoted by $\bar{\lambda}, \bar{Q}$. For fixed auxiliary fields, the Gaussian Lagrangian $L(\bar{\lambda}, \bar{Q})$ describes noninteracting z particles. $\bar{\lambda}, \bar{Q}$ are determined by the saddle point equations[2]

$$\left.\frac{\delta S}{\delta \lambda_i}\right|_{\bar{\lambda}, \bar{Q}} = \langle n_i \rangle - S = 0, \tag{17.1}$$

$$\left.\frac{\delta S}{\delta Q_{ij}}\right|_{\bar{\lambda}, \bar{Q}} = \left.\frac{\delta S}{\delta Q^*_{ij}}\right|_{\bar{\lambda}, \bar{Q}} = 0, \tag{17.2}$$

which are commonly called the *"mean field equations."* $\langle n \rangle$ denotes a mean field expectation value using the saddle point Lagrangian $L(\bar{\lambda}, \bar{Q})$. In mean field theories described in Chapter 18, $\bar{\lambda}, \bar{Q}$ are static and uniform fields, i.e., one has only two variational parameters whose inputs are the lattice structure, spin, and temperature.

Loosely speaking, the physical meaning of mean field theory is the "best" quadratic Lagrangian that approximates the energy and correlations of the Heisenberg model. However, we emphasize that *it is not a variational theory*. The mean field Hamiltonian and ground state admix unphysical wave functions which violate the local constraints. Also, the mean field excitations ("quasiparticles") violate the constraints by including density

---

[1]See (10.25).

[2]There are cases where the saddle points of the fields Q and Q^* are not complex conjugates; see Auerbach and Larson.

fluctuations of Schwinger bosons or constrained fermions. Constraint conserving excitations are composites of even numbers of quasiparticles. The local constraints are reintroduced into the theory through the dynamical fluctuations of the field λ. This statement will be made precise in Section 17.3.1.

Q_{ij} are effective fields which couple to the quasiparticles. The physical interpretation of these fields is given by the mean field equation (17.2),

$$0 = \frac{\delta S}{\delta Q_{ij}^*}\bigg|_{\bar{\lambda},\bar{Q}} \Rightarrow Q_{ij} = \begin{cases} \frac{J}{N}\langle\mathcal{F}_{ij}\rangle & \text{FM-B} \\ \frac{J}{N}\langle\mathcal{A}_{ij}\rangle & \text{AFM-B} \\ \frac{J}{N}\langle\mathcal{D}_{ij}\rangle & \text{AFM-F} \end{cases} . \qquad (17.3)$$

Thus Q_{ij} depend self-consistently on the expectation values of the bond operators.

17.1   Fluctuations and Gauge Fields

We restrict the discussion of this section to the Schwinger boson models FM-B and AFM-B. There is a semiclassical interpretation for the fluctuations of the Hubbard–Stratonovich fields Q_{ij} in terms of physical spin correlations. This is elucidated by the matrix elements of $\mathcal{F}_{ij}$ and $\mathcal{A}_{ij}$ for $N = 2$, between spin coherent states defined in Section 7.3. The ferromagnetic bond operator yields

$$\begin{aligned} Q_{ij}^F &\Rightarrow \frac{J}{2}\langle\hat{\Omega}_i|_S\langle\hat{\Omega}_j|_{S-\frac{1}{2}}\,\mathcal{F}_{ij}\,|\hat{\Omega}_i\rangle_{S-\frac{1}{2}}|\hat{\Omega}_j\rangle_S \\ &= JS\,e^{i(\chi_i-\chi_j)/2}\,(u_i^*u_j + v_i^*v_j) \\ &= JS\exp\left(-\frac{i}{2}\psi_{ij}^F\right)\left|(1+\hat{\Omega}_i\cdot\hat{\Omega}_j)/2\right|^{\frac{1}{2}} , \qquad (17.4) \end{aligned}$$

where $|\hat{\Omega}(\theta,\phi)\rangle_S$ and $u(\theta,\phi),v(\theta,\phi)$ are defined in (7.18) and (7.14), respectively. The phase ψ^F is given by (7.19). By (17.4) we see that Q is maximized for ferromagnetically correlated spins $\hat{\Omega}_i \approx \hat{\Omega}_j$.

For smoothly varying configurations, ψ_{ij}^F can be expanded to linear order in the spin deviations (see (7.19)),

$$\begin{aligned} \psi_{i,i+\delta_\mu}^F &= -\cos(\theta_i)(\phi_i - \phi_{i+\delta_\mu}) + \chi_i - \chi_{i+\delta_\mu} + \mathcal{O}\left(|\delta_\mu|^2\right) \\ &\simeq \mathbf{A}(\hat{\Omega}_i)\cdot\partial_\mu\hat{\Omega}_i|\delta_\mu| . \qquad (17.5) \end{aligned}$$

$\{\delta_\mu\}$ are the nearest neighbor vectors on the lattice. $\mathbf{A}$ is the monopole vector potential defined in (10.18), and χ_i's are arbitrary gauge choices.

A continuous *gauge field* is defined by

$$A_\mu(\mathbf{x}_i,\tau) \equiv \frac{1}{2}\mathbf{A}(\hat{\Omega}_i)\cdot\partial_\mu\hat{\Omega}_i , \qquad \mu = 0,1,\ldots,d , \qquad (17.6)$$

where $\partial_0 = \partial_\tau$. The gauge invariant ("electromagnetic") field describes chiral spin correlations

$$F_{\mu\nu} = \partial_\mu A_\nu - \partial_\nu A_\mu = \frac{1}{2}\partial_\mu\hat{\Omega} \times \partial_\nu\hat{\Omega} \cdot \hat{\Omega}. \tag{17.7}$$

A finite value $F_{\mu\nu} \neq 0$ implies *noncoplanarity* of the local spin correlations in the $\mu - \nu$ plane. That is to say: the spins in a finite area on the $\mu - \nu$ plane subtend a finite area on the unit sphere.

For the antiferromagnetic case, the bond operator yields

$$Q_{ij}^N \Rightarrow \frac{J}{2}\langle\hat{\Omega}_i|_{S-\frac{1}{2}}\langle\hat{\Omega}_j|_{S-\frac{1}{2}} A_{ij} |\hat{\Omega}_i,\hat{\Omega}_j\rangle_S = JSe^{-i(\chi_j+\chi_i)/2}(u_iv_j - v_iu_j)$$

$$= JS\exp\left(-\frac{i}{2}\psi_{ij}^N\right) \left|(1 - \hat{\Omega}_i \cdot \hat{\Omega}_j)/2\right|^{\frac{1}{2}}. \tag{17.8}$$

Q^N is maximized for antiferromagnetically (Néel) correlated spins $\hat{\Omega}_i \approx -\hat{\Omega}_j$. The Néel field is defined as

$$\hat{n}(x_i) \approx \frac{1}{2}\left[\eta_i\hat{\Omega}_i + (2d)^{-1}\sum_{\delta_\mu}\eta_{i+\delta_\mu}\hat{\Omega}_{i+\delta_\mu}\right], \qquad \eta_i = \begin{cases} 1 & i \in A \\ -1 & i \in B \end{cases}. \tag{17.9}$$

For antiferromagnetically correlated configurations, $\hat{n}$ is approximately a unit vector field parametrized by $(\tilde{\theta}, \tilde{\phi})$. The bond phases ψ_{ij}^N are approximately given by

$$\psi_{i,i+\delta_\mu}^N \simeq -\eta_i\cos(\tilde{\theta}_i)(\tilde{\phi}_i - \tilde{\phi}_{i+\delta_\mu}) + (\chi_i - \chi_{i+\delta_\mu})$$

$$\simeq \eta_i\mathbf{A}(\hat{n}) \cdot \partial_\mu\hat{n}|\delta_\mu|. \tag{17.10}$$

The *Néel gauge field* A^N is defined as in (17.6):

$$A_\mu^N(x_i,\tau) \equiv \frac{\eta_i}{2}\mathbf{A}(\hat{n}) \cdot \partial_\mu\hat{n}, \tag{17.11}$$

where again $\partial_0 = \partial_\tau$. The Néel "electromagnetic" field describes chiral fluctuations of the staggered magnetization,

$$F_{\mu\nu}^N = \partial_\mu A_\nu^N - \partial_\nu A_\mu^N = \frac{1}{2}\partial_\mu\hat{n} \times \partial_\nu\hat{n} \cdot \hat{n}. \tag{17.12}$$

In one dimension, the "electric field" F_{01}^N is

$$F_{01}^N = \frac{1}{2}\partial_\tau\hat{n} \times \partial_x\hat{n} \cdot \hat{n}, \tag{17.13}$$

where the right-hand side is proportional to the "topological density" in the (x,τ) plane.[3]

The association between the phases of Q_{ij}^N and chiral Néel correlations has been useful for understanding the correspondence between the large N and the semiclassical continuum approaches.[4]

---

[3]See Section 15.1, and Exercises 2 and 3 of Chapter 13.
[4]See Read and Sachdev, bibliography of Chapter 18.

17.2 $1/N$ Expansion Diagrams

We denote the fluctuations about the saddle points by

$$r_\alpha = [i\lambda_i(\tau) - \bar{\lambda}, \ \operatorname{Re} Q_{ij}(\tau) - \operatorname{Re} \bar{\mathbf{Q}}, \ \operatorname{Im} Q_{ij}(\tau) - \operatorname{Im} \bar{Q}], \qquad (17.14)$$

where α includes field, space, and time indices. By (17.1) and (17.2), the Taylor expansion of the action about the saddle point has no linear terms,

$$\mathcal{S} = \mathcal{S}^{(0)} + \frac{1}{2}\mathcal{S}^{(2)}_{\alpha\alpha'} r_\alpha r_{\alpha'} + \mathcal{S}^{int}, \qquad (17.15)$$

where summation over repeated indices is assumed. $\mathcal{S}^{int}$ includes third- and higher-order terms:

$$\mathcal{S}^{int} = \sum_{n=3}^{\infty} \frac{1}{n!}\mathcal{S}^{(n)}_{\alpha_1 \dots \alpha_n} \, r_{\alpha_1} \cdots r_{\alpha_n}. \qquad (17.16)$$

$\mathcal{S}^{(n)}$ are determined as follows. The action is written as

$$
\begin{aligned}
\mathcal{S} &= \mathcal{S}_0 + \frac{\eta}{N}\operatorname{Tr}\ln\left(1 + G_0 \sum_\alpha r_\alpha \hat{v}_\alpha\right), \\
\mathcal{S}_0 &= \frac{\eta}{N}\operatorname{Tr}\ln G_0^{-1} + \int_0^\beta d\tau \left(\sum_{\langle ij \rangle} \frac{|Q_{ij}|^2}{J} - iS \sum_i \lambda_i\right), \\
G_0 &= \hat{G}(\bar{Q}, \bar{Q}^*, \bar{\lambda}),
\end{aligned}
\qquad (17.17)
$$

where $\hat{G}$ was defined in (16.37) and (16.38). G_0 is the mean field Green's function. The matrices $\hat{v}_\alpha$ are "internal vertices," which conserve m-flavor and connect the field r_α to two G_0's. Using the identity

$$\operatorname{Tr}\ln(1 + A) = -\sum_{n=1}^{\infty} \frac{(-1)^n}{n}\operatorname{Tr}(A)^n, \qquad (17.18)$$

we obtain an explicit expression for

$$\mathcal{S}^{(n)}_{\alpha_1 \dots \alpha_n} = \frac{(-1)^{n+1}}{n}\frac{\eta}{N}\sum_P \eta \operatorname{Tr}(\hat{v}_{\alpha_1} G_0 \cdots \hat{v}_{\alpha_n} G_0). \qquad (17.19)$$

$\sum_P$ is a sum over all permutations of $(\alpha_1, \dots \alpha_n)$. The trace sums over space, time, and m-flavors. $\mathcal{S}^{(n)}$ are diagramatically depicted as "loops" of mean field Green's functions of the same m index with n internal vertices, see Fig. 17.1.

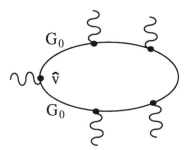

FIGURE 17.1. $\mathcal{S}^{(5)}$ – a loop with five internal vertices.

The *"RPA propagator"* is the inverse of the quadratic form

$$D_{\alpha,\alpha'} = \left(\frac{1}{2}\mathcal{S}^{(2)}\right)^{-1}_{\alpha,\alpha'} = (\Pi_0 - \Pi)^{-1}_{\alpha,\alpha'}, \qquad (17.20)$$

where

$$\Pi_{\alpha,\alpha'} = \frac{\eta}{N}\text{Tr}\left(\hat{v}_\alpha G_0 \hat{v}_{\alpha'} G_0\right) . \qquad (17.21)$$

Π_0 is diagonal in momentum and Matsubara space. In the auxiliary field representation $(\text{Re}\,Q, \text{Im}\,Q, \lambda)$, it is given by

$$\Pi_0 = \begin{pmatrix} \frac{1}{J} & 0 & 0 \\ 0 & \frac{1}{J} & 0 \\ 0 & 0 & 0 \end{pmatrix}. \qquad (17.22)$$

The term "RPA" (random phase approximation) is commonly used to describe Gaussian fluctuations about mean field theory. $\det D$ should always be calculated in order to verify that

$$\text{Re}\det D > 0. \qquad (17.23)$$

Zeros and negative values of $\text{Re}\det D$ are *dangerous*, since they invalidate the mean field theory as a legitimate saddle point.

By expanding the prefactors and exponential in (16.40) in powers of r_α, the correlation function is given by the sums

$$\tilde{R}^{mm'}(1,2) = \tilde{R}^I(1,2) + \delta_{mm'}\tilde{R}^{II}(1,2),$$

$$\tilde{R}^I = Z^{-1}\int \mathcal{D}r \left(\sum_{n=0}^{\infty} -\frac{1}{n!}\mathcal{S}^{(n+2)}_{1,2,\alpha_1,\alpha_n}r_{\alpha_1}\cdots r_{\alpha_n}\right)$$

$$\times \sum_{L=0}^{\infty}\frac{(-N)^L}{L!}\left(\sum_{n=3}^{\infty}\frac{1}{n!}\mathcal{S}^{(n)}_{\alpha_1\dots\alpha_n}r_{\alpha_1}\cdots r_{\alpha_n}\right)^L \exp\left(-\frac{N}{2}\mathcal{S}^{(2)}_{\alpha\beta}r_\alpha r_\beta\right),$$

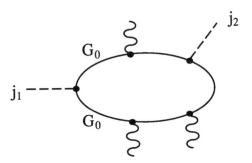

FIGURE 17.2. A Loop of five vertices, two of which are external.

$$\tilde{R}^{II} = Z^{-1} \int \mathcal{D}r \left(\sum_{n=0}^{\infty} \frac{1}{n!} S^{(n+1)}_{1,\alpha_1,\alpha_n} r_{\alpha_1} \cdots r_{\alpha_n} \right) \left(\sum_{n=0}^{\infty} \frac{1}{n!} S^{(n+1)}_{2,\alpha_1,\alpha_n} r_{\alpha_1} \cdots r_{\alpha_n} \right)$$

$$\times \sum_{L=0}^{\infty} \frac{(-N)^L}{L!} \left(\sum_{n=3}^{\infty} \frac{1}{n!} S^{(n)}_{\alpha_1 \ldots \alpha_n} r_{\alpha_1} \cdots r_{\alpha_n} \right)^L \exp\left(-\frac{N}{2} S^{(2)}_{\alpha\beta} r_\alpha r_\beta \right),$$

$$(17.24)$$

where we have disregarded the constant $-N S^{(0)}$ in the action. We define the *external loops* which contain one or two current vertices (see Fig. 17.2)

$$\frac{\delta}{\delta j_m(1)} S^{(n)}_{\alpha_1,\alpha_n} = \frac{1}{N} S^{(n+1)}_{1,\alpha_1,\alpha_n},$$

$$\frac{\delta^2}{\delta j_m(1) \delta j_m(2)} S^{(n)}_{\alpha_1,\alpha_n} = S^{(n+2)}_{1,2,\alpha_1,\alpha_n}. \qquad (17.25)$$

The integrals in (17.24) are sums of multidimensional Gaussian integrals. The property of Gaussian integrals is that an integral over an even number of fields can be written as a sum over pairwise *"contractions"*

$$\overbrace{r_1 \cdots r_{2k}} = \sum_{i_1, \ldots i_k} \overbrace{r_1 r_{i_1}} \cdots \overbrace{r_k r_{i_k}}, \qquad (17.26)$$

where the contraction of two variables yields

$$\overbrace{r_1 r_2} \equiv Z^{-1} \int \mathcal{D}r \, r_{\alpha i \tau} r_{\alpha' i' \tau'} \exp\left(-\frac{N}{2} r_\alpha D^{-1}_{\alpha,\beta} r_\beta \right)$$

$$= \frac{1}{N} D_{1,2}. \qquad (17.27)$$

The propagator is depicted as a wavy line which connects two vertices in Fig. 17.3. Given a certain number of loops, diagrams are constructed where

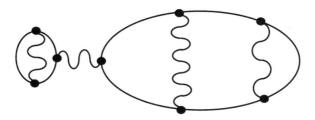

FIGURE 17.3. Diagrams with contracted internal fields.

the vertices are connected by D's in all possible ways. A disconnected part of the diagram contains loops that are not connected to the external loops (17.25). For the correlation function, we need only consider the connected diagrams, since the disconnected parts cancel the normalization factor Z^{-1}. This is the *"linked cluster theorem."* Thus, calculating any particular diagram involves multiplying all possible loops with vertices contracted by propagators and summing over internal indices. An important condition is that the internal loops (which do not contain an external vertex) must have at least *three vertices*. External loops might have any number (including zero) of internal vertices. The order in $1/N$ of any particular diagram is given by

$$\left(\frac{1}{N}\right)^{P-L}, \tag{17.28}$$

where L is the number of internal loops, and P is the number of propagators. After summing over all diagrams at order $1/N^p$, we obtain $\tilde{R}^{(p)}$, which are the coefficients of the series

$$\tilde{R}^{mm'}(1,2) \sim \sum_{p=0}^{\infty} N^{-p} \tilde{R}^{mm'(p)}(1,2). \tag{17.29}$$

The $\sim$ denotes an asymptotic series. Equation (17.29) may not converge, and it may miss essentially singular terms such as $e^{-N(\cdot)}$.

17.3 Sum Rules

The diagramatic expansion of the $1/N$ series has a special structure that allows us to obtain certain identities, or sum rules, which hold at each order in the $1/N$ series. By the very definition of D, the constraint of no charge fluctuations is enforced order by order. We shall see how this works below for both Schwinger bosons and constrained fermions. In addition, we derive a sum rule for the local SU(N) spin fluctuations, for all N.

17.3.1   ABSENCE OF CHARGE FLUCTUATIONS

Here we shall demonstrate that the constraint is imposed exactly to each order in $1/N$. In other words, the density fluctuations vanish identically after all diagrams of a given order are summed, yielding

$$\langle\langle n_1 \mathcal{A}\rangle\rangle = NS \langle\langle \mathcal{A}\rangle\rangle, \tag{17.30}$$

for any operator $\mathcal{A}$, where $\langle\langle \cdot \rangle\rangle$ is the exact expectation value at any temperature.

It is instructive to see how (17.30) is derived by the diagramatic expansion. The Lagrangian in the presence of the corresponding source current contains the term

$$(j_1 + i\lambda_1)\left(\sum_m z_{1m}^* z_{1m}\right) - i\lambda_1 NS. \tag{17.31}$$

The external vertex with respect to j_1 is identical to the internal vertex of the constraint field λ_1. By differentiating $S[j]$ twice using (17.25), we obtain

$$S_{1,\alpha}^{(2)} = -i\Pi_{\lambda_1,\alpha}, \tag{17.32}$$

and by (17.20) and (17.22), we obtain

$$\sum_\alpha \Pi_{\lambda_1,\alpha} D_{\alpha,\alpha'} = -(I_\lambda)_{\lambda_1,\alpha'} + (\Pi_0 D)_{\lambda_1,\alpha'} = -(I^\lambda)_{\lambda_1,\alpha'}, \tag{17.33}$$

where I^λ projects onto the λ field sector. In order to evaluate (17.30) diagrammatically, we consider the connected diagrams that contain the external loop $S^{(n+1)}$, $n \geq 0$. Let us first consider all possible diagrams with $n \geq 1$. We define a *"tail"* of a diagram as the combination of a propagator attached in series to a loop $\Pi_{\lambda_1,r}$ which was defined in (17.21). All diagrams can be separated into two classes: ones with a tail, and ones without a tail. It is easy to identify for each diagram without a tail, say, $\Gamma(n_1, \mathcal{A})$, a counterterm $\bar{\Gamma}(n_1, \mathcal{A})$ by attaching a tail to the j_1 vertex, as depicted in Fig. 17.4. By (17.28) they are both of the same order in $1/N$, since they have the same number of loops minus propagators. Using (17.33) we prove that the term and the counterterm must precisely cancel:

$$
\begin{aligned}
\bar{\Gamma}(n_1, \mathcal{A}) &= N\sum_{\alpha\alpha'} \Pi_{\lambda_1\alpha} D_{\alpha,\alpha'} \Gamma(n_l, \mathcal{A}) \\
&= \sum_{\alpha'}\left[-\delta_{\lambda_1,\alpha'} + \sum_\alpha (\Pi_0)_{\lambda_1,\alpha} D_{\alpha,\alpha'}\right]\Gamma(\alpha', \mathcal{A}) \\
&= -\Gamma(n_1, \mathcal{A}).
\end{aligned}
\tag{17.34}
$$

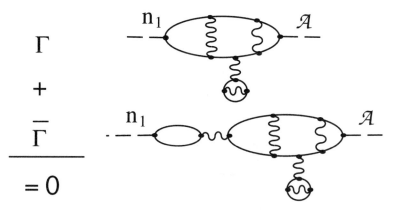

FIGURE 17.4. A charge fluctuation diagram and its counterterm.

Thus, to any order of $1/N$ the counterterms cancel the connected terms in the charge fluctuation diagrams. The only terms that survive are the disconnected diagrams with n_1 on the loop $\mathcal{S}^{(1)}$. The diagram rules exclude a counterterm to the loop $N\mathcal{S}^{(1)}$ since there can be no internal loop with one or two vertices. The saddle point condition (17.1) implies that

$$\langle n_1 \rangle = NS, \tag{17.35}$$

which completes the proof of (17.30). Q.E.D.

17.3.2 ON-SITE SPIN FLUCTUATIONS

Here we shall evaluate the *on-site* spin fluctuations. For SU(2), we are familiar with the "spin square" operator $\mathbf{S}^2$, which when projected to the S sector yields $S(S+1)$ times the identity matrix. The SU(N) generators

$$\hat{S}_i^{mm'} = a_{im}^\dagger a_{im'} \tag{17.36}$$

define the on-site fluctuations as

$$S^{mm'} \equiv \langle\!\langle \hat{S}_i^{mm'} \hat{S}_i^{m'm} \rangle\!\rangle, \tag{17.37}$$

where $\langle\!\langle \cdot \rangle\!\rangle$ is the exact expectation value, at any temperature. Using the constraint, it is easy to determine the expectation value, of the "spin square" operator,

$$\begin{aligned}
\sum_{mm'} S^{mm'} &= \langle\!\langle \sum_{mm'} [n_{im'}(1 + \eta n_{im}) - \eta \delta_{mm'} n_{im'}] \rangle\!\rangle \\
&= N^2 S(1 + \eta S) - \eta NS. \tag{17.38}
\end{aligned}$$

(For the constrained fermions case, we have used the identity $n_{im}^2 = n_{im}$.)

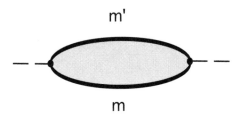

m'

m

FIGURE 17.5. $S^{m \neq m'}$ — the sum of all propagator irreducible diagrams.

Since the path integral is $SU(N)$ invariant, we can determine the individual correlations $S^{mm'}$ as follows. First, we find the relation between diagonal fluctuations S^{mm} and the off-diagonal fluctuations $S^{m \neq m'}$. In a rotationally symmetric system, these quantities cannot depend on m or m'.

By (17.24), $S^{m \neq m'}$ is given by the path integral $\tilde{R}^I$. Since internal vertices conserve the m index, all diagrams in $S^{m \neq m'}$ are connected and cannot be separated into two parts by cutting any number of propagator lines. We denote these diagrams as *propagator irreducible* (PI) and depict them in Fig. 17.5. All PI diagrams that constitute $S^{m \neq m'}$ also appear in the connected part of the *diagonal* correlations S_c^{mm}, i.e.,

$$\text{if} \quad \Gamma^{PI} \in S^{m \neq m'}, \quad \text{then} \quad \Gamma^{PI} \in S_c^{mm}. \tag{17.39}$$

However, for each Γ^{pi}, S_c^{mm} also contains a counterterm $\bar{\Gamma}^{PI}$ which has a tail attached to it on one end, see Fig. 17.6. Using (17.33), the term and its counterpart add up to

$$\Gamma^{PI}(i,i) + \bar{\Gamma}^{PI}(i,i) = \Gamma^{PI}(i,i) + \sum_{\alpha\alpha'} \bar{\Gamma}_\alpha^{PI}(i,\alpha) D_{\alpha,\alpha'} \Pi_{\alpha'\lambda_i}$$

$$= (1 - 1/N)\Gamma^{PI}(i,i). \tag{17.40}$$

In addition to PI diagrams and their counterterms, there are contributions to S_c^{mm} that have the two external vertices on different loops. These are *propagator reducible* (PR) diagrams. However, it is easy to show that all PR diagrams cancel exactly with counterterms, as shown in Fig. 17.7. This is done by adding a tail to one of the loops and seeing that the counterterm is of the same order in $1/N$ due to the additional m'-summation for the loop that became internal. This proves that all the diagrams that do not cancel are PI. Using (17.38), we obtain the following important identity:

$$S_c^{mm}(i,i) = \left(1 - \frac{1}{N}\right) S^{m \neq m'}(i,i). \tag{17.41}$$

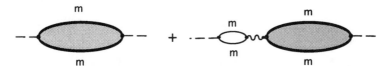

FIGURE 17.6. S_c^{mm} — propagator irreducible diagrams and their counterterms.

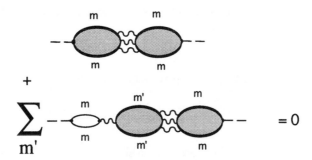

FIGURE 17.7. S_c^{mm} — cancellation of propagator reducible diagrams.

By separating the connected and disconnected parts of S^{mm} and using (17.41), we obtain the on-site fluctuations as

$$\sum_{mm'} S^{m,m'} = \sum_{m \neq m'} S^{m \neq m'}(i,i) + \sum_m \left[S_c^{mm}(i,i) + \langle\!\langle n_{im} \rangle\!\rangle^2 \right]$$

$$= \left[N(N-1) + N(1 - \frac{1}{N}) \right] S^{m \neq m'}(i,i) + NS^2.$$

$$(17.42)$$

Equating (17.38) and (17.42), we obtain the desired identity

$$S^{m \neq m'}(1,1) = \frac{N}{N+\eta} S(1 + \eta S). \qquad (17.43)$$

For $N = 2$, (17.43) reduces to the known result:

$$\langle S_i^+ S_i^- \rangle = \begin{cases} \frac{2}{3}S(S+1) & \text{Schwinger bosons} \\ \frac{1}{2} & \text{constrained fermions} \end{cases}. \tag{17.44}$$

(17.43) can be used to check diagramatic calculations of spin correlations at any order in $1/N$. In momentum and Matsubara frequency representation, the sum of all the diagrams of order $(1/N)^p$ must obey the *total moment sum rule*:

$$(\beta N)^{-1} \sum_{\mathbf{k},n} S^{m \neq m'(p)}(\mathbf{k}, \omega_n) = \left(\frac{1}{N}\right)^p (-\eta)^p S(1+\eta S), \quad p = 0, 1, 2, \dots . \tag{17.45}$$

17.4   Exercises

1. Evaluate the spin correlations for the one-dimensional half filled Fermi gas

$$|\Psi^{FG}\rangle = \prod_{|k| < \pi/2} c_{k\uparrow}^\dagger c_{k\downarrow}^\dagger |0\rangle. \tag{17.46}$$

Hint: Prove the following identity and evaluate it:

$$S^{+-}(q) = \frac{1}{N} \sum_k \langle c_{k\uparrow}^\dagger c_{k+q\downarrow} c_{k+q\downarrow}^\dagger c_{k\uparrow} \rangle. \tag{17.47}$$

2. By summing over q, evaluate the on-site fluctuations $\langle S_i^+ S_i^- \rangle$. Compare the result to the sum rule (17.44). Explain the discrepancy.

Bibliography

The effects of constraints in the large N expansion could be learned from the large N expansion of correlations in Gutzwiller projected wave functions, see bibliography of Chapter 8.

Mean field theories where the saddle points of the Hubbard–Stratonovich fields Q and Q^* are not complex conjugates are discussed in

• A. Auerbach and B.E. Larson, Phys. Rev. B **43**, 7800 (1991).

18

Schwinger Bosons Mean Field Theory

In this chapter, we review the large-N Schwinger bosons mean field theory (SBMFT) for the case of the Heisenberg ferromagnet (FM-B) and antiferromagnet (AFM-B) in one dimension and on the square lattice. The SBMFT (in contrast to spin wave theory of Chapter 11) maintains an explicit SU(2) symmetry of the effective Hamiltonian.[1] We shall see that the SBMFT recovers the main results of the continuum approaches (except for Θ term effects) described in Chapter 15.

In the following, N is held as an independent parameter. We set $N \to 2$ when evaluating spin correlations of the physical Heisenberg model.

The SBMFT is given by replacing the auxiliary fields in the action of (16.39) by *static and uniform* saddle point parameters:

$$N\mathcal{S}\left[Q_{ij}(\tau), \lambda_i(\tau)\right] \to N\mathcal{S}_0(Q, -i\lambda) = \beta F^{MF}(Q, \lambda). \tag{18.1}$$

F^{MF} is the mean field free energy, which can be written as

$$F^{MF}(Q, \lambda) = -\beta^{-1} \ln \text{Tr}_{im} \left[\exp\left(-\beta H^{MF}[Q, \lambda]\right)\right], \tag{18.2}$$

where H^{MF} is the mean field Hamiltonian of N decoupled boson flavors.

18.1 The Case of the Ferromagnet

Using (18.1) and (16.36), the mean field Hamiltonian for the FM-B in the absence of external current sources is as follows:

$$
\begin{aligned}
H^{MF} &= \sum_{im} \lambda a_{im}^\dagger a_{im} - Q \sum_{\langle ij \rangle, m} \left(a_{im}^\dagger a_{jm} + a_{jm}^\dagger a_{im}\right) \\
&\qquad + N\mathcal{N}\frac{zQ^2}{2J} - N\mathcal{N}S\lambda \\
&= \sum_{\mathbf{k}, m} \epsilon_{\mathbf{k}} a_{\mathbf{k}m}^\dagger a_{\mathbf{k}m} + N\mathcal{N}\frac{zQ^2}{2J} - N\mathcal{N}S\lambda. \tag{18.3}
\end{aligned}
$$

---

[1] A modified spin wave theory which is similar to the SBMFT but breaks spin rotational symmetry was developed by M. Takahashi.

$\mathcal{N}$ and z are the number of lattice points and number of nearest neighbors, respectively, $\mathbf{k}$ is the lattice momentum, and

$$
\begin{aligned}
\epsilon_{\mathbf{k}} &= \lambda - zQ\gamma_{\mathbf{k}} , \\
\gamma_{\mathbf{k}} &= \frac{1}{z}\sum_{\hat{\eta}} e^{i\mathbf{k}\hat{\eta}} .
\end{aligned}
\tag{18.4}
$$

η are the nearest neighbor vectors.

The mean field free energy is

$$
F^{MF} = N\sum_{\mathbf{k}} \ln\left(1 - e^{-\epsilon_{\mathbf{k}}/T}\right) + N\mathcal{N}\frac{zQ^2}{2J} - N\mathcal{N}S\lambda.
\tag{18.5}
$$

The saddle point equations (17.1) and (17.2) are obtained by minimizing F with respect to λ and Q:

$$
\frac{1}{\mathcal{N}}\sum_{\mathbf{k}} n_{\mathbf{k}} = S ,
\tag{18.6}
$$

$$
\frac{1}{\mathcal{N}}\sum_{\mathbf{k}} n_{\mathbf{k}}\gamma_{\mathbf{k}} = Q/J,
\tag{18.7}
$$

where

$$
n_{\mathbf{k}} = \frac{1}{e^{\epsilon_{\mathbf{k}}/T} - 1}.
\tag{18.8}
$$

Equations (18.6) and (18.7) uniquely determine the mean field parameters $Q(T, S)$ and $\lambda(T, S)$.

It is convenient to use the more physical parametrization of the mean field variables,

$$
\begin{aligned}
\epsilon_{\mathbf{k}} &= zQ(1 - \gamma_{\mathbf{k}} + \frac{1}{4z}\kappa^2), \\
\kappa^2 &= 4z(\lambda/zQ - 1).
\end{aligned}
\tag{18.9}
$$

zQ, κ describe the bandwidth and the inverse correlation length, respectively. In Fig. 18.1 the mean field dispersion for the ferromagnet in one dimension is shown.

By subtracting (18.7) from (18.6), one obtains an explicit expression for Q:

$$
Q = JS - \frac{J}{\mathcal{N}}\sum_{\mathbf{k}} n_{\mathbf{k}}(1 - \gamma_{\mathbf{k}}).
\tag{18.10}
$$

Although at small arguments $n_{\mathbf{k}} \sim (\beta\epsilon_{\mathbf{k}})^{-1}$, the integrand on the right-hand side is bounded at $\mathbf{k} \to 0$, and the sum has a regular power series in T. At low temperatures $T \ll zQ$, one obtains

$$
Q = JS + \mathcal{O}(T/JS).
\tag{18.11}
$$

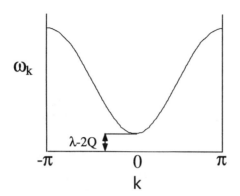

FIGURE 18.1. Mean field dispersion for the one-dimensional ferromagnet.

At finite temperatures, $\kappa > 0$ and $\epsilon_{\mathbf{k}}$ has a gap of $zQ\kappa^2/8$ at $\mathbf{k} = 0$. As we take $T \to 0$, and $\mathcal{N} \to \infty$ (the *thermodynamic limit*), the Bose function vanishes for all $\mathbf{k} \neq 0$. In order to satisfy the constraint equation (18.6), there must be a macroscopic occupation at $\mathbf{k} = 0$:

$$S = \frac{1}{\mathcal{N}} \sum_{\mathbf{k}} n_{\mathbf{k}} \sim \frac{1}{\mathcal{N}} n_0 = \frac{4T}{\mathcal{N}JS\kappa^2} + \mathcal{O}(T^2), \qquad (18.12)$$

which implies that

$$\lim_{T \to 0} \kappa = \sqrt{\frac{4T}{\mathcal{N}JS^2}}. \qquad (18.13)$$

In order to study the spin correlations, we restrict ourselves to $N = 2$, and introduce a magnetic field in the z direction, which couples as the term

$$H^{MF} \quad \to \quad H^{MF} - h \sum_i S_i^z$$

$$= \quad H^{MF} - h \sum_{i, s = \pm \frac{1}{2}} s a_{is}^\dagger a_{is}. \qquad (18.14)$$

The field splits the degeneracy between the two Schwinger boson dispersions

$$\epsilon_{\mathbf{k}m} = \epsilon_{\mathbf{k}} - hs. \qquad (18.15)$$

The constraint equation yields

$$S \quad = \quad \frac{1}{2} \lim_{T \to 0} \mathcal{N}^{-1} \sum_{\mathbf{k}} \left[n(\epsilon_{\mathbf{k}, +\frac{1}{2}}) + n(\epsilon_{\mathbf{k}, -\frac{1}{2}}) \right]$$

$$\to \quad \frac{1}{2} \mathcal{N}^{-1} n(\epsilon_{0, +\frac{1}{2}}), \qquad (18.16)$$

and the magnetization is given by[2]

$$m_0 = \frac{1}{2}\mathcal{N}^{-1}\lim_{h \to 0+}\sum_{\mathbf{k}}\left[n(\epsilon_{\mathbf{k},+\frac{1}{2}}) - n(\epsilon_{\mathbf{k},-\frac{1}{2}})\right]$$

$$= \frac{1}{2}\mathcal{N}^{-1}n(\epsilon_{0,+\frac{1}{2}}) = S. \tag{18.17}$$

Thus, under an infinitesimal magnetic field the Bose condensation at $T=0$ occurs at $\mathbf{k} = 0, s = +\frac{1}{2}$. The Bose condensate is complete in the sense that all the available Schwinger boson density ($2S$ per site) is accumulated in that single mode. The order parameter for the Schwinger boson condensate is

$$\langle a_{\mathbf{k}s}\rangle = \langle a_{\mathbf{k}s}^{\dagger}\rangle = \sqrt{2\mathcal{N}m_0}\delta_{s,\frac{1}{2}}\delta_{\mathbf{k},0}. \tag{18.18}$$

This order parameter violates the constraint (i.e., it connects states of different spins). Nevertheless, it has a precise meaning within the mean field theory and the $1/N$ expansion.[3]

The condensation of all bosons into the lowest state translates to the spin state

$$\Psi^{MF} = \prod_i (a_{i,+\frac{1}{2}}^{\dagger})^{2S}|0\rangle = |S,S,\ldots\rangle, \tag{18.19}$$

which agrees with the exact ground state as given by Corollary 5.6 of Marshall's Theorem; see (5.37).

We note that the SBMFT dispersion $\epsilon_{\mathbf{k}}$ reduces at $T \to 0$ to ferromagnetic spin waves, which vanish as (see (11.49)):

$$\lim_{T \to 0}\epsilon_{\mathbf{k}} \to \omega_{\mathbf{k}} = zJS(1 - \gamma_{\mathbf{k}})$$

$$\sim JS|\mathbf{k}|^2 + \mathcal{O}\left(|\mathbf{k}|^4\right). \tag{18.20}$$

The vanishing of the Schwinger boson energy at $\mathbf{k} = 0$ is expected by Goldstone's Theorem; see Section 9.2.

The susceptibility is given by

$$\chi^{MF} = -\mathcal{N}^{-1}\frac{d^2F(h)}{d^2h} = -\frac{1}{4\mathcal{N}}\frac{d^2F(h)}{d^2\lambda}$$

$$= \frac{1}{4\mathcal{N}T}\sum_{\mathbf{k}}n_{\mathbf{k}}(n_{\mathbf{k}} + 1). \tag{18.21}$$

The mean field spin correlations are given by the free boson expression

$$\langle S_i^+ S_j^-\rangle_{MF} = \langle a_{i,+\frac{1}{2}}^{\dagger}a_{i,-\frac{1}{2}}a_{j,-\frac{1}{2}}^{\dagger}a_{j,+\frac{1}{2}}\rangle$$

---

[2]See (6.3).
[3]See Hirsch and Tang.

$$= |R_{ij}|^2 + S\delta_{ij} \, ,$$
$$R_{ij} = \frac{1}{N} \sum_{\mathbf{k}} n_{\mathbf{k}} e^{i\mathbf{k} \cdot \mathbf{x}_{ij}}, \tag{18.22}$$

where we have used the identity (18.6). By (18.6) and (18.22) the local spin fluctuations are

$$\langle S_i^+ S_i^- \rangle_{MF} = \frac{1}{N} \sum_{\mathbf{k},\mathbf{k}'} n_{\mathbf{k}} (n_{\mathbf{k}'} + 1) = S(S+1). \tag{18.23}$$

For SU(2), the exact result is, of course, $\frac{2}{3}S(S+1)$. By the sum rule (17.43), we verify that (18.23) recovers the correct $(1/N)^0$ contribution to the fluctuations.

18.1.1 One Dimension

In one dimension at low temperature $T \ll JS$, the sum in the constraint equation (18.6) is dominated by the region of small energies and momenta. For small k,

$$\epsilon \approx JS \left(\frac{1}{4}\kappa^2 + k^2 + \ldots \right) \, ,$$
$$n(\epsilon) \approx T/\epsilon. \tag{18.24}$$

At low temperatures, we replace the upper momentum cutoff by infinity. This yields the leading-order temperature dependence of the constraint sum

$$N^{-1} \sum_{\mathbf{k}} n_{\mathbf{k}} \sim \frac{T}{JS\pi} \int_0^\infty dk \, \frac{1}{\frac{1}{4}\kappa^2 + k^2} + \mathcal{O}(T^2)$$
$$\sim T/(JS\kappa) \approx S. \tag{18.25}$$

Thus, we obtain

$$\kappa \sim T/(JS^2) + \mathcal{O}(T^2). \tag{18.26}$$

The low-temperature susceptibility is also dominated by the small-momenta part of the integral. By expanding the term (18.21) in powers of k, we can calculate its leading temperature dependence:

$$\chi^{MF} = \frac{1}{2T} \int_{-\pi}^{\pi} \frac{dk}{2\pi} \, n_k(n_k + 1)$$
$$\sim \frac{JS^4}{T^2}. \tag{18.27}$$

The exact susceptibility χ of the $S = \frac{1}{2}$ Heisenberg ferromagnet has been calculated by Takahashi using Bethe's solution. He finds that

$$\chi^{S=\frac{1}{2}} \sim \frac{J}{24} T^{-2} \sim \frac{2}{3}\chi^{MF} \tag{18.28}$$

This discrepancy of 2/3 was previously found in the local spin fluctuations (18.23). It is surprising, however, to find the same factor for a *long-range* spin correlation function such as χ.

R_{ij} of (18.22) can be evaluated at large distances by the behavior of the integrand at low momenta:

$$R_{ij} \sim \frac{T}{JS\pi} \int_0^\infty dk \, \frac{e^{ik|i-j|}}{\frac{1}{4}\kappa^2 + k^2} = S \, e^{-\frac{1}{2}\kappa|i-j|}, \qquad (18.29)$$

where we have used (18.26). At low temperatures and large distances, using (18.6) we obtain

$$\langle S_i^z S_j^z \rangle_{MF} \sim \frac{1}{2} S^2 e^{-|i-j|/\xi}, \qquad (18.30)$$

which establishes that the parameter κ is the inverse correlation length

$$\xi = \kappa^{-1} \sim JS^2/T. \qquad (18.31)$$

This result agrees with the exact correlation length of the *classical* Heisenberg ferromagnet in one dimension evaluated by M. Fisher.

18.1.2 Two Dimensions

We consider the square lattice, where $z = 4$. A useful function for evaluating momenta sums is

$$\rho(\gamma) = \mathcal{N}^{-1} \sum_{\mathbf{k}} \delta(\gamma - \gamma_{\mathbf{k}}) = \frac{2}{\pi^2} K(1 - \gamma^2), \qquad (18.32)$$

which describes the density of states of the tight binding hopping matrix on the square lattice, where

$$\gamma_{\mathbf{k}} = \frac{1}{2}(\cos k_x + \cos k_y). \qquad (18.33)$$

$K(m)$ is the *"Complete Elliptic Integral of the First Kind"*[4] defined by

$$K(m) = \int_0^1 dt \, \left[(1 - t^2)(1 - mt^2) \right]^{-\frac{1}{2}}$$

$$\approx \begin{cases} \theta(m) \left[\frac{\pi}{2} + \frac{\pi}{8} m + \mathcal{O}(m)^2 \right] & m \ll 1 \\ \frac{1}{2} \ln \left[16/(1 - m) \right] & |1 - m| \ll 1 \end{cases}. \qquad (18.34)$$

The important features of $\rho(\gamma)$ are the step function discontinuities at $\gamma = \pm 1$ and the logarithmic divergence near $\gamma_{\mathbf{k}} = 0$. This function is plotted in Fig. 18.2. Since the dominant low-temperature correlations depend on

---

[4]See Abramowitz and Stegun, p. 590.

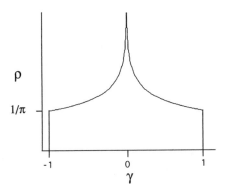

FIGURE 18.2. Density of states for the tight binding dispersion $\gamma_\mathbf{k}$ on the square lattice.

$\rho(\gamma)$ at $\gamma \approx 1$, we simplify further calculations by replacing

$$\rho(\gamma) \;\rightarrow\; \bar{\rho} \;=\; \frac{1}{\pi}, \qquad \gamma \in [1-\pi, 1],$$

$$\int_{1-\pi}^{1} d\gamma\, \bar{\rho}(\gamma) \;=\; 1, \tag{18.35}$$

where we have extended the domain of γ to keep a unit total density. With this quite harmless approximation, it is easy to solve the constraint equation:

$$\mathcal{N}^{-1}\sum_\mathbf{k} n_\mathbf{k} \;\approx\; \frac{T}{4JS\pi}\int_{1-\pi}^{1} d\gamma\, \frac{1}{4(1-\gamma)+\kappa^2/4}$$

$$= \frac{T}{JS\pi}\ln(16\pi/\kappa^2)+\mathcal{O}(T^2) \;=\; S. \tag{18.36}$$

Inverting this equation and using (18.11), we obtain the leading order in temperature solution:

$$\kappa \;\approx\; \sqrt{16\pi}\exp\left(\frac{-2\pi JS^2}{T}\right). \tag{18.37}$$

The spin correlations at large distances can be evaluated using the Ornstein–Zernicke approximation[5] to R_{ij},

$$R_{ij} \;\approx\; \frac{T}{JS}\int \frac{d^2\mathbf{k}}{(2\pi)^2}\frac{e^{i\mathbf{k}\mathbf{x}_{ij}}}{\frac{1}{4}\kappa^2+k^2}$$

$$\propto\; (|\mathbf{x}_{ij}|/\xi)^{-\frac{1}{2}}\exp\left(-|\mathbf{x}_{ij}|\kappa/2\right)\left(1+\frac{2}{|\mathbf{x}_{ij}|\kappa}\cdots\right), \tag{18.38}$$

---

[5]See (13.48).

where $\mathbf{x}_{ij} = \mathbf{x}_i - \mathbf{x}_j$. By (18.22), we find that

$$\langle S_i^z S_j^z \rangle_{MF} \propto \frac{\xi}{|\mathbf{x}_{ij}|} e^{-|\mathbf{x}_{ij}|/\xi} , \qquad (18.39)$$

where

$$\xi = \kappa^{-1} , \qquad (18.40)$$

i.e., κ^{-1} is the spin correlation length. We therefore conclude that the correlation length increases exponentially with decreasing temperature and that there is no long-range order at any finite $T > 0$, in accordance with Mermin and Wagner's theorem.

Recall that we have seen in Section 13.1 that the *classical Heisenberg ferromagnet* is described at long distances and low temperatures by the nonlinear sigma model with coupling constant:

$$f = \frac{T}{JS^2}. \qquad (18.41)$$

We find that (18.37) recovers the renormalization group result to one loop order (13.55) and the large N approximation to the CP^{N-1} model (14.26).

18.2   The Case of the Antiferromagnet

By (16.31) and (18.2), the mean field Hamiltonian for the AFM-B representation is

$$
\begin{aligned}
H^{MF} &= \sum_{i,m} \lambda a_{im}^\dagger a_{im} + Q \sum_{\langle ij \rangle, m} \left(a_{im}^\dagger a_{jm}^\dagger + a_{im} a_{jm} \right) \\
&\quad + N\mathcal{N}\frac{zQ^2}{2J} - N\mathcal{N}S\lambda \\
&= \sum_{\mathbf{k}m} \left[\lambda a_{\mathbf{k}m}^\dagger a_{\mathbf{k}m} + \frac{1}{2} zQ\gamma_{\mathbf{k}} \left(a_{\mathbf{k}m}^\dagger a_{-\mathbf{k}m}^\dagger + a_{\mathbf{k}m} a_{-\mathbf{k}m} \right) \right] \\
&\quad + N\mathcal{N}\frac{zQ^2}{2J} - N\mathcal{N}S\lambda.
\end{aligned} \qquad (18.42)
$$

For $N = 2$, the SBMFT Hamiltonian resembles the Holstein–Primakoff spin wave Hamiltonian H_1 given in (11.60), except that here two Schwinger boson flavors replace the single Holstein–Primakoff boson. In close analogy to the spin wave problem, H^{MF} can be diagonalized by a canonical *Bogoliubov* transformation

$$\alpha_{\mathbf{k}m} = \cosh\theta_{\mathbf{k}} a_{\mathbf{k}m} - \sinh\theta_{\mathbf{k}} a_{-\mathbf{k}m}^\dagger, \qquad (18.43)$$

or inversely,

$$a_{\mathbf{k}m} = \cosh\theta_{\mathbf{k}} \alpha_{\mathbf{k}m} + \sinh\theta_{\mathbf{k}} \alpha_{-\mathbf{k}m}^\dagger. \qquad (18.44)$$

By inserting (18.44) in (18.42), one obtains a normal diagonal Hamiltonian in terms of the α bosons,

$$
\begin{aligned}
H^{MF} &= \frac{1}{2}\sum_{\mathbf{k}m}\Big[(\lambda\cosh 2\theta_{\mathbf{k}} + zQ\gamma_{\mathbf{k}}\sinh 2\theta_{\mathbf{k}})(\alpha^{\dagger}_{\mathbf{k}m}\alpha_{\mathbf{k}m} + \alpha_{\mathbf{k}m}\alpha^{\dagger}_{\mathbf{k}m}) \\
&\quad + (\lambda\sinh 2\theta_{\mathbf{k}} + zQ\gamma_{\mathbf{k}}\cosh 2\theta_{\mathbf{k}})(\alpha^{\dagger}_{\mathbf{k}m}\alpha^{\dagger}_{-\mathbf{k}m} + \alpha_{\mathbf{k}m}\alpha_{-\mathbf{k}m})\Big] \\
&\quad + NN\frac{zQ^2}{2J} - NN\left(S + \frac{1}{2}\right)\lambda.
\end{aligned}
\tag{18.45}
$$

Now, we choose $\theta_{\mathbf{k}}$ so that the anomalous terms $\alpha^{\dagger}\alpha^{\dagger}$ and $\alpha\alpha$ vanish. This amounts to the following condition on $\theta_{\mathbf{k}}$:

$$
\tanh 2\theta_{\mathbf{k}} = -\frac{zQ\gamma_{\mathbf{k}}}{\lambda}.
\tag{18.46}
$$

Having solved for $\theta_{\mathbf{k}}$, we can substitute the hyperbolic functions in (18.45) by rational functions of the right-hand side of (18.46). This yields a normal and diagonal Hamiltonian:

$$
\begin{aligned}
H^{MF} &= \sum_{\mathbf{k}m}\omega_{\mathbf{k}}\left(\alpha^{\dagger}_{\mathbf{k}m}\alpha_{\mathbf{k}m} + \frac{1}{2}\right) + NN\frac{zQ^2}{2J} - NN\left(S + \frac{1}{2}\right)\lambda, \\
\omega_{\mathbf{k}} &= \sqrt{\lambda^2 - (zQ\gamma_{\mathbf{k}})^2}.
\end{aligned}
\tag{18.47}
$$

The mean field free energy is given by

$$
F^{MF} = \beta^{-1}\sum_{\mathbf{k}m}\ln\left[2\sinh\left(\frac{\beta\omega_{\mathbf{k}}}{2}\right)\right] - NN\left(S + \frac{1}{2}\right)\lambda + NN\frac{zQ^2}{2J}.
\tag{18.48}
$$

The mean field equations are given by differentiating (18.48) with respect to λ and Q:

$$
\frac{1}{N}\sum_{\mathbf{k}}\frac{\lambda}{\sqrt{\lambda^2 - (zQ\gamma_{\mathbf{k}})^2}}\left(n_{\mathbf{k}} + \frac{1}{2}\right) = S + \frac{1}{2},
\tag{18.49}
$$

$$
\frac{1}{N}\sum_{\mathbf{k}}\frac{z^2\gamma_{\mathbf{k}}^2 Q}{\sqrt{\lambda^2 - (zQ\gamma_{\mathbf{k}})^2}}\left(n_{\mathbf{k}} + \frac{1}{2}\right) = \frac{zQ}{J}.
\tag{18.50}
$$

The mean field ground state Ψ^{MF} is the vacuum of all α's,

$$
\alpha_{\mathbf{k},m}\Psi^{MF}_0 = 0, \qquad \forall\mathbf{k},m.
\tag{18.51}
$$

Using (18.43), one can write Ψ^{MF} explicitly in terms of the original Schwinger bosons as

$$
\Psi^{MF} = C\exp\left[\frac{1}{2}\sum_{ij}u_{ij}\left(\sum_{m}a^{\dagger}_{im}a^{\dagger}_{jm}\right)\right]|0\rangle,
$$

$$u_{ij} = \frac{1}{N} \sum_{\mathbf{k}} e^{i\mathbf{k}\mathbf{x}_{ij}} \tanh\theta_{\mathbf{k}}. \qquad (18.52)$$

For $N = 2$, using the *unrotated* operators $a^\dagger, b^\dagger$, Ψ^{MF} is the Schwinger bosons mean field state (8.5),

$$\Psi_{N=2}^{MF} = |\hat{u}\rangle = \exp\left[\sum_{i\in A, j\in B} u_{ij}\left(a_i^\dagger b_j^\dagger - b_i^\dagger a_j^\dagger\right)\right] |0\rangle. \qquad (18.53)$$

Ψ^{MF} contains many configurations with occupations different from $2S$ and is therefore not a pure spin state. As shown in Chapter 8, under Gutzwiller projection it reduces to a valence bond state. Since

$$\tanh(\theta_{\mathbf{k}+\vec{\pi}}) = -\tanh(\theta_{\mathbf{k}}) , \qquad (18.54)$$

where $\vec{\pi} = (\pi, \pi, \ldots)$, the bond parameters u_{ij} only connect sublattice A to B. Furthermore, one can verify that for the nearest neighbor model above, $u_{ij} \geq 0$, and therefore the valence bond states obey Marshall's sign as defined in (5.13).

Although Ψ^{MF} are manifestly rotationally invariant, they may or may not have long-range antiferromagnetic order. This depends on the long-distance decay of u_{ij}. As we shall see, the SBMFT ground state for the nearest neighbor model is disordered in one dimension and has long-range order in two dimensions.

For further calculations, it is convenient to introduce the parametrizations:

$$\omega_{\mathbf{k}} \equiv c\sqrt{(\kappa/2)^2 + \frac{z}{2}(1 - \gamma_{\mathbf{k}}^2)},$$

$$c \equiv Q\sqrt{2z},$$

$$\kappa \equiv \frac{2}{c}\sqrt{\lambda^2 - (zQ)^2},$$

$$t = \frac{T}{zQ}. \qquad (18.55)$$

c, κ, t describe the spin wave velocity, the inverse correlation length, and the dimensionless temperature, respectively. In Fig. 18.3 the dispersion for the one-dimensional antiferromagnet is drawn. By (18.48), we see that near the zone center and zone corner the mean field dispersions are those of free massive relativistic bosons,

$$\omega_{\mathbf{k}}^{\gamma} \approx c\sqrt{(\kappa/2)^2 + |\mathbf{k} - \mathbf{k}_\gamma|^2} , \qquad \mathbf{k}_\gamma = 0, \vec{\pi}. \qquad (18.56)$$

When the gap (or "mass" $c\kappa/2$) vanishes, $\omega_{\mathbf{k}}^{\alpha}$ are Goldstone modes which reduce to dispersions of antiferromagnetic spin waves (11.65).

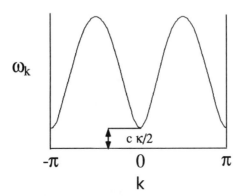

FIGURE 18.3. Mean field dispersion for the one-dimensional antiferromagnet.

The spin correlation function is given by inserting the α operators instead of a's in (16.40) using (18.44). This yields

$$
\begin{aligned}
S^{MF}(\mathbf{q}) &= \frac{1}{N}\langle S_{\mathbf{q}}^{+} S_{-\mathbf{q}}^{-}\rangle_{MF} \\
&= \frac{1}{N}\sum_{\mathbf{k}}\left\{\cosh\left[2\left(\theta_{\mathbf{k}}+\theta_{\mathbf{k}+\mathbf{q}+\bar{\pi}}\right)\right]\right. \\
&\qquad \left. \times\left(n_{\mathbf{k}}+\frac{1}{2}\right)\left(n_{\mathbf{k}+\mathbf{q}+\bar{\pi}}+\frac{1}{2}\right)-\frac{1}{4}\right\}.
\end{aligned}
\tag{18.57}
$$

Using (18.49), we confirm the large N limit of the sum rule (17.43),

$$
\frac{1}{N}\sum_{\mathbf{q}}S^{MF}(\mathbf{q}) = S(S+1).
\tag{18.58}
$$

For $N=2$, the mean field sum rule exceeds the exact result by a familiar factor of $\frac{3}{2}$.

The spatial dependence of the spin correlations at $\mathbf{x}_{ij} = \mathbf{x}_i - \mathbf{x}_j$ is given by

$$
S^{MF}(\mathbf{x}_{ij}) = |f(\mathbf{x}_{ij})|^2 - |g(\mathbf{x}_{ij})|^2 - \frac{1}{4}\delta_{ij},
\tag{18.59}
$$

where at low temperatures and long distances,

$$
f(\mathbf{x}_{ij}) = N^{-1}\sum_{\mathbf{k}}\frac{\left(n_{\mathbf{k}}+\frac{1}{2}\right)e^{i\mathbf{k}\mathbf{x}_{ij}}}{\sqrt{1-\gamma_{\mathbf{k}}^{2}[\kappa^2/(2z)+1]^{-1}}}
$$

$$\approx\ 2zt\left(1+e^{i\vec{\pi}\mathbf{X}_{ij}}\right)\int\frac{d^d\mathbf{k}}{(2\pi)^d}\frac{e^{i\mathbf{k}\mathbf{x}_{ij}}}{(\kappa/2)^2+|\mathbf{k}|^2}$$

$$\propto\ \left(1+e^{i\vec{\pi}\mathbf{X}_{ij}}\right)(\mathbf{x}_{ij}|/\xi)^{-(d-1)/2}\exp\left(-|\mathbf{x}_{ij}|\kappa/2\right),\quad(18.60)$$

where we have used (13.48) for the long-distance behavior at small κ and low temperatures. Similarly,

$$g(\mathbf{x}_{ij})\ =\ \mathcal{N}^{-1}\sum_{\mathbf{k}}\gamma_{\mathbf{k}}\frac{\left(n_{\mathbf{k}}+\frac{1}{2}\right)e^{i\mathbf{k}\mathbf{x}_{ij}}}{\sqrt{1-\gamma_{\mathbf{k}}^2[\kappa^2/(2z)+1]^{-1}}}$$

$$\propto\ \left(1-e^{i\vec{\pi}\mathbf{X}_{ij}}\right)(|\mathbf{x}_{ij}|/\xi)^{-(d-1)/2}\exp\left(-|\mathbf{x}_{ij}|\kappa/2\right).\quad(18.61)$$

Thus, for $\kappa>0$,

$$S^{MF}(\mathbf{x}_{ij})\ \propto\ e^{i\vec{\pi}\mathbf{X}_{ij}}\left(\frac{\xi}{|\mathbf{x}_{ij}|}\right)^{d-1}\exp\left(-|\mathbf{x}_{ij}|/\xi\right),\quad(18.62)$$

where the correlation length is $\xi=\kappa^{-1}$.

The uniform susceptibility is obtained directly from (18.57) using the identity

$$\chi_0^{MF}\ =\ \frac{1}{2T}S^{MF}(\mathbf{q}=0)\ =\ \frac{1}{T}\sum_{\mathbf{k}}n_{\mathbf{k}}(n_{\mathbf{k}}+1).\quad(18.63)$$

18.2.1 LONG-RANGE ANTIFERROMAGNETIC ORDER

In the absence of any magnetic fields, the ground state is a singlet. In two dimensions and higher, we have seen in Chapter 11 that the nearest neighbor antiferromagnet has long-range order for all $S\geq\frac{1}{2}$. To investigate the possibility of spontaneously broken symmetry, we introduce an infinitesimal ordering field h which couples to the staggered magnetization. We restrict ourselves to $N=2$; thus

$$H^{MF}\ \rightarrow\ H^{MF}-h\sum_i S_i^z$$

$$=\ H^{MF}-h\sum_{i,s=-\frac{1}{2},\frac{1}{2}}sa_{is}^\dagger a_{is}\ ,\quad(18.64)$$

where we recall that the Schwinger bosons are defined using sublattice rotated spin directions in (16.11). Equation (18.64) can be diagonalized using spin dependent transformation angles $\theta_{\mathbf{k},s}$. Repeating the steps leading to (18.47), we obtain

$$\omega_{\mathbf{k},s}\ =\ \sqrt{(\lambda-sh)^2-(zQ\gamma_{\mathbf{k}})^2}.\quad(18.65)$$

The spontaneous staggered magnetization is given by the limit

$$m_0 = \lim_{h\to 0^+} m(h) ,$$

$$m(h) = \lim_{\mathcal{N}\to\infty} \frac{1}{\mathcal{N}} \sum_{i,s} s \, \langle a_{is}^\dagger a_{is} \rangle_h$$

$$= \lim_{\mathcal{N}\to\infty} \frac{1}{\mathcal{N}} \sum_{\mathbf{k}s} \frac{\lambda - sh}{\sqrt{(\lambda - sh)^2 - (zQ\gamma_\mathbf{k})^2}} \left[n(\omega_{\mathbf{k},s}) + \frac{1}{2} \right].$$

$$(18.66)$$

The constraint equation (18.49) is

$$\frac{1}{2\mathcal{N}} \sum_{\mathbf{k},s} \frac{\lambda - sh}{\sqrt{(\lambda - sh)^2 - (zQ\gamma_\mathbf{k})^2}} \left[n(\omega_{\mathbf{k},s}) + \frac{1}{2} \right] = S + \frac{1}{2}. \qquad (18.67)$$

Again, we parametrize the dispersions in terms of $c, \kappa, \tilde{h}$,

$$\omega_{\mathbf{k},s} \equiv c\sqrt{\kappa^2/4 - \left(s - \frac{1}{2}\right) z\tilde{h} + \frac{z}{2}(1 - \gamma_\mathbf{k}^2)} + \mathcal{O}(\tilde{h}^2),$$

$$c = \sqrt{2zQ},$$

$$\kappa = \frac{2}{c}\sqrt{(\lambda - \frac{1}{2}h)^2 - (zQ)^2},$$

$$\tilde{h} = h\frac{\lambda}{(zQ)^2}. \qquad (18.68)$$

If

$$\lim_{\mathcal{N}\to\infty} \kappa > 0, \qquad (18.69)$$

then both summands in (18.66), with $s = \pm\frac{1}{2}$, are continuous at $h = 0$, and therefore

$$\lim_{h\to 0^+} m(h,\kappa) = 0 , \qquad (18.70)$$

i.e., no spontaneous symmetry breaking. On the other hand, if

$$\kappa = \mathcal{O}(\mathcal{N})^{-1} \qquad (18.71)$$

the $s = +\frac{1}{2}$ summand at $\mathbf{k} = 0, \vec{\pi}$ contributes a term of order $\mathcal{N}$. This yields a *macroscopic contribution* (i.e., order one) to the staggered magnetization (18.66). Since it also represents a macroscopic contribution to the Schwinger bosons density (18.67), we can say that there is Bose condensation of the $s = +\frac{1}{2}$ bosons at $\mathbf{k} = 0, \vec{\pi}$ (see discussion after (18.18)). The order parameter for the condensate is thus

$$\langle a_{\mathbf{k}s} \rangle = \langle a_{\mathbf{k}s}^\dagger \rangle = \sqrt{\frac{\mathcal{N}m_0}{2}} \delta_{s,\frac{1}{2}} \left(\delta_{\mathbf{k},0} + \delta_{\mathbf{k},\vec{\pi}} \right). \qquad (18.72)$$

To evaluate $m_0 = \lim_{h \to 0} m(h)$, we subtract (18.66) from (18.67), eliminate the diverging $s = \frac{1}{2}$ summand, and obtain

$$
\begin{aligned}
m_0 = {} & S + \frac{1}{2} \\
& - \lim_{h \to 0^+} \lim_{\mathcal{N} \to \infty} \mathcal{N}^{-1} \sum_{\mathbf{k}} \frac{2 + 2\tilde{h} + \kappa^2/4}{\sqrt{2\tilde{h} + \kappa^2/4 + 2(1 - \gamma_{\mathbf{k}}^2)}} \left[n(\omega_{\mathbf{k}\frac{1}{2}}) + \frac{1}{2} \right].
\end{aligned}
\tag{18.73}
$$

By keeping $h > 0$, we maintain a gap in the spectrum and in the denominator of the summand. Thus, we are allowed to replace (18.73) by an integral in the thermodynamic limit, and at $T = 0$ we can set $n(\omega_{\mathbf{k},+\frac{1}{2}}) = 0$ and $\kappa = 0$. This yields

$$
m_0 = S + \frac{1}{2} - \frac{1}{2} \int \frac{d^d \mathbf{k}}{(2\pi)^d} \frac{1}{\sqrt{1 - \gamma_{\mathbf{k}}^2}}.
\tag{18.74}
$$

The integral yields for cubic lattices in d dimensions the numerical results

$$
m_0(d) = \begin{cases} 0 & d = 1 \\ S - 0.19660 & d = 2 \\ S - 0.078 & d = 3 \end{cases}.
\tag{18.75}
$$

Notice that, in contrast to the ferromagnetic case (18.17), the ordered moment is always *less than* the classical value S. This is due to the quantum zero-point motion, which has its origin in the noncommutability of the Hamiltonian and the staggered magnetization. The SBMFT results for m_0 agree with low-order spin wave theory given in (11.69).

18.2.2 ONE DIMENSION

At zero temperature, we set $n_{\mathbf{k}} = 0$ and expand (18.49) as

$$
\begin{aligned}
S + \frac{1}{2} &= \frac{1}{2} \int_{-\pi}^{\pi} \frac{dk}{2\pi} \frac{1}{\sqrt{1 - \frac{1}{1 + \kappa^2/8} \cos^2(k)}} \\
&= \frac{1}{\pi} K \left(\frac{1}{1 + \kappa^2/8} \right) \\
&\sim \frac{1}{\pi} \ln \left(\frac{8\sqrt{2}}{\kappa} \right) + \mathcal{O}(\kappa),
\end{aligned}
\tag{18.76}
$$

which results in

$$
\kappa \approx \sqrt{32} \exp \left[-\pi \left(S + \frac{1}{2} \right) \right].
\tag{18.77}
$$

In (18.75), we found that there cannot be long-range order in the SBMFT ground state of the one-dimensional antiferromagnet. Since κ decreases exponentially with S, we can neglect κ as we neglect higher-order corrections in S^{-1}. By subtracting (18.50) from (18.49) one obtains

$$
\begin{aligned}
c &= J\left(S+\frac{1}{2}-\frac{1}{2}\int_{-\pi}^{\pi}\frac{dk}{2\pi}\frac{\sin^2 k}{\sqrt{\kappa^2/8+\sin^2 k}}\right) \\
&\approx J\left[S+\frac{1}{2}-\frac{2}{\pi}+\mathcal{O}(\kappa,S^{-1})\right].
\end{aligned}
\tag{18.78}
$$

While c does not differ drastically from its classical value $c = JS$ (see (11.41)), the ground state correlations decay exponentially. The correlation length κ^{-1} as given by (18.77) agrees with the correlation length given by the continuum approach in (15.14).

The mean field excitations are not physical excitations of the Heisenberg model (for example, they include constraint violating charge fluctuations). Nevertheless, the gap in their spectrum $c\kappa/2$ is consistent with the existence of Haldane's gap for one-dimensional integer spin chains. The physical magnon spectrum can be deduced from the peaks in Im $S(\mathbf{q},\omega)$. Since spin one excitations involve at least two Schwinger bosons, the physical magnons have a gap of

$$
\Delta = c\kappa.
\tag{18.79}
$$

Thus, the SBMFT recovers Haldane's continuum results in the absence of a Θ term (see Chapter 15).

One must beware that the SBMFT fails for half-odd integer spin chains. The effects of the Θ term (15.3), which destroys Haldane's gap, are apparently absent in the mean field theory. The SBMFT has a nondegenerate ground state in violation of Lieb, Schultz, and Mattis' theorem of Section 5.2 for half-odd integer spins. Read and Sachdev overcame this problem by introducing the Θ term into the large N theory. They developed a continuum gauge theory for the Q, λ fluctuations (see bibliography).

18.2.3   Two Dimensions

In two dimensions, we expect no long-range order at finite temperatures due to Mermin and Wagner's theorem. Indeed, we shall find that, at low temperatures, the mean field equations yield a finite value for the inverse correlation length $\kappa(T,S)$.

The mean field equations can be solved to obtain $\kappa(T,S), c(T,S)$. It is convenient to use the fact that $\omega_{\mathbf{k}} = \omega(\gamma_{\mathbf{k}})$ and replace

$$
\frac{1}{N}\sum_{\mathbf{k}}F(\omega_{\mathbf{k}}) \to 2\int_0^1 d\gamma\, \rho(\gamma)F\left[\omega(\gamma)\right],
\tag{18.80}
$$

where for the square lattice

$$\rho(\gamma) = \frac{2}{\pi^2} K(1 - \gamma^2). \tag{18.81}$$

It turns out that at temperatures above $T_{max} > 0.91 J$, the mean field equations have no nontrivial $(Q \neq 0)$ solution.[6] This reflects a failure of the SBMFT to describe the disordered phase at high temperatures, where nearest neighbor correlations are destroyed. At temperatures where the correlation length is large, i.e.,

$$\kappa \ll t \ll 1, \tag{18.82}$$

the constraint equation can be expanded following Takahashi:

$$
\begin{aligned}
S + \frac{1}{2} &= \frac{1}{2} \int_{-1}^{1} d\gamma \, \rho(\gamma) \left(1 + \kappa^2/8 - \gamma^2\right)^{-\frac{1}{2}} \coth\left[(1 + \kappa^2/8 - \gamma^2)^{\frac{1}{2}}/2t\right] \\
&= \frac{t}{\pi} \left[\log\left(\frac{32}{\kappa^2}\right) - \log\left(\frac{2}{t}\right) \right] \\
&\quad + \frac{1}{\pi^2} \int_{-1}^{1} d\gamma (1 - \gamma^2)^{-\frac{1}{2}} K(1 - \gamma^2) + \mathcal{O}(t, \kappa).
\end{aligned} \tag{18.83}
$$

Similarly, by subtracting (18.50) from (18.49) and expanding the integrals to low order in κ, t, we obtain

$$
\begin{aligned}
S + \frac{1}{2} &- \frac{4Q}{J} \\
&\geq \frac{1}{2} \int_{-1}^{1} d\gamma \rho(\gamma) \left(1 + \kappa^2/8 - \gamma^2\right)^{\frac{1}{2}} \coth\left[(1 + \kappa^2/8 - \gamma^2)^{\frac{1}{2}}/2t\right] \\
&\approx \frac{1}{\pi^2} \int_{-1}^{1} d\gamma (1 - \gamma^2)^{\frac{1}{2}} K(1 - \gamma^2) + \mathcal{O}(t^3, \kappa).
\end{aligned} \tag{18.84}
$$

By (18.74) and (18.75) the ordered moment is given by

$$m_0 = S + \frac{1}{2} - \frac{1}{\pi^2} \int_{-1}^{1} d\gamma (1 - \gamma^2)^{-\frac{1}{2}} K(1 - \gamma^2) = S - 0.19660. \tag{18.85}$$

The spin wave velocity at zero temperature is

$$c = \sqrt{8} J S Z_c, \tag{18.86}$$

where the spin wave velocity renormalization factor is given using (18.84),

$$
\begin{aligned}
Z_c &= 1 + S^{-1} \left[\frac{1}{2} - \frac{1}{\pi^2} \int_{-1}^{1} d\gamma (1 - \gamma^2)^{\frac{1}{2}} K(1 - \gamma^2) \right] \\
&= 1 + 0.078974/S.
\end{aligned} \tag{18.87}
$$

---

[6]See Arovas and Auerbach.

The asymptotic spin correlations are given by (18.62), where the correlation length is $\xi = \kappa^{-1}$. By inverting (18.83) and using (18.84), we obtain the temperature dependent correlation length

$$\xi = \frac{\sqrt{2}JSZ_c}{T} \exp\left[-\frac{2\pi Z_c JSm_0}{T}\right][1 + \mathcal{O}(t^2)] . \qquad (18.88)$$

In comparing (18.88) to the continuum approximation value[7] (15.17), we can determine the renormalized stiffness constant $\rho_s(S)$ for the square lattice model as follows:

$$\rho_s = \lim_{T \to 0} \frac{T}{2\pi} \log(\xi) = JSm_0 Z_c, \qquad (18.89)$$

where m_0 and Z_c are given by (18.85) and (18.87), respectively. It properly recovers the classical value $\rho_s \to JS^2$ in the limit of large S.

18.3 Exercises

1. Using the explicit solutions for Q and λ at zero tempearture, prove that the ground state energies for the FM-B and the AFM-B mean field theories are given by

$$E^{MF} = \lim_{T \to 0} F^{MF} = -N\mathcal{N}\frac{zQ^2}{2J}. \qquad (18.90)$$

Adding the constant $z\mathcal{N}JS^2/2$, which was dropped from the Hamiltonian (see (16.7)), show that the correct ferromagnetic ground state energy for $N = 2$ is recovered.

2. Prove that a staggered magnetic field term (18.64) for the antiferromagnetic SBMFT results in the dispersions $\omega_{\mathbf{k}_s}$ of (18.65). *Hint: Find the zeros of the determinant of the quadratic matrix*

$$L = \sum_s (z^\dagger_{\mathbf{k}_s}, \bar{z}_{-\mathbf{k}_s}) \begin{pmatrix} \omega - \lambda + \frac{1}{2}h_s & z\gamma_{\mathbf{k}}Q \\ z\gamma_{\mathbf{k}}Q & -\omega - \lambda + \frac{1}{2}h_s \end{pmatrix} \begin{pmatrix} z_{\mathbf{k}_s} \\ \bar{z}^\dagger_{-\mathbf{k}_s} \end{pmatrix}, \qquad (18.91)$$

where z and $\bar{z}$ are coherent state variables of sublattices A and B, respectively.

3. For the antiferromagnetic model, add a *uniform* magnetic field to H^{MF} of (18.42) as follows:

$$H^{MF} \to H^{MF} - h \sum_{is=\pm\frac{1}{2}} se^{i\vec{\pi}\mathbf{x}_i} a^\dagger_{is}a_{is}. \qquad (18.92)$$

Allow the constraint field to have uniform and staggered components:

$$\lambda_i = \lambda + \exp(i\vec{\pi}\mathbf{x}_i)\lambda_s. \qquad (18.93)$$

---

[7]See Chakravarty, Halperin, and Nelson, bibliography of Chapter 15.

Show that the mean field ground state energy is minimized for $\lambda_s = -\frac{1}{2}h$. Discuss how λ_s can be interpreted as a uniform precession at angular frequency $\omega = -\frac{1}{2}h$ of all spins in the xy plane.

4. Using the results of the previous exercise, derive the uniform susceptibility χ_0^{MF} of (18.63) as a second derivative of the free energy with respect to h_0. Note that it is necessary to keep the temperature finite before taking the thermodynamic limit. Why?

Bibliography

The Schwinger bosons mean field theory was introduced in

- D. P. Arovas and A. Auerbach, Phys. Rev. B **38**, 316 (1988).

The SBMFT dynamical spin correlations for the square lattice antiferromagnet are given in

- A. Auerbach and D. P. Arovas, Phys. Rev. Lett. **61**, 617 (1988).

The correlation length of the classical Heisenberg ferromagnet in one dimension was evaluated by

- M. Fisher, Am. J. Phys. **32**, 343 (1964).

Elliptic functions are defined in

- M. Abramowitz and I.A. Stegun, *Handbook of Mathematical Functions* (Dover, 1965), Chapter 17 .

Modified spin wave theory, which includes analytic solutions to the SBMFT equations, is found in

- M. Takahashi, Prog. Theor. Phys. Suppl. **87**, 233 (1986);
- M. Takahashi, Phys. Rev. B **36**, 3791 (1987);
- M. Takahashi, Phys. Rev. B **40**, 2494 (1989).

The connection between long-range magnetic order and Schwinger boson condensation is explained in

- J.E. Hirsch and S. Tang, Phys. Rev. B **39**, 2850 (1989);
- M. Raykin and A. Auerbach, Phys. Rev. Lett. **70**, 3808 (1993).

The continuum mapping of a large family of SU(N) Heisenberg models and the recovery of topological Berry phase terms was done by

- N. Read and S. Sachdev, Nucl. Phys. B **316**, 609 (1989);
- N. Read and S. Sachdev, Phys. Rev. Lett. 62, 1694 (1989).

19

The Semiclassical Theory of the $t - J$ Model

The discovery of high-temperature superconductivity in layered copper oxides (e.g., $La_{2-x}Sr_xCuO_4$) has spurred intense investigations of the quasi-two-dimensional Hubbard model near half filling. The basic problem is to understand the evolution of the antiferromagnetic Mott-type insulator such as La_2CuO_4 into a metal or superconductor at hole concentrations of x=5–20%.

As shown in Section 3.2, the low-lying excitations of the Hubbard model at large U can be described by the $t - J$ Hamiltonian (3.29)

$$\mathcal{H}^{t-J} = P_s \left(\mathcal{T} + \mathcal{H}^{QHM} + \mathcal{J}' \right) P_s, \tag{19.1}$$

where P_s projects the Hilbert space onto the subspace of zero and one electron per site, and

$$
\begin{aligned}
\mathcal{T} &= -t \sum_{\langle ij \rangle s} \left(c_{is}^\dagger c_{js} + c_{js}^\dagger c_{is} \right), \\
\mathcal{H}^{QHM} &= J \sum_{\langle ij \rangle} \left(\mathbf{S}_i \cdot \mathbf{S}_j - \frac{n_i n_j}{4} \right), \\
\mathcal{J}' &= -\frac{1}{2U} \sum_{\langle i,j,k \rangle}^{i \neq k} t_{ij} t_{jk} \left(\sum_s (c_{is}^\dagger c_{ks} \rho_j) - c_i^\dagger \vec{\sigma} c_k \cdot c_j^\dagger \vec{\sigma} c_j \right).
\end{aligned}
\tag{19.2}
$$

$\langle ij \rangle$ are nearest neighbor bonds, and $\langle i, j, k \rangle$ denotes a sum over sites i, j, k, which are triads of nearest neighbors. At half filling, $n_i \rightarrow 1$, the hopping terms $\mathcal{T}$ and $\mathcal{J}'$ vanish in the projected subspace, and the $t - J$ model reduces to the quantum Heisenberg antiferromagnet of spin half.

Away from half filling, the $t - J$ model describes a system of interacting spins and mobile holes, also called the *"doped antiferromagnet."*

Since large U/t implies $t >> J$, the t-hopping term is not small. Moreover, hopping of holes between opposite sublattices disturbs the antiferromagnetic correlations. This presents a fundamental theoretical difficulty: the Heisenberg model, which we believe we understand quite well, is strongly perturbed by addition of mobile holes. This is especially manifested in the extreme Nagaoka limit[1] $J = 0$, where the ground state of a single hole is totally ferromagnetic!

When choosing an approximation scheme for the $t - J$ model, it would be useful to identify a control parameter with which to estimate our errors. Since we are interested in low doping concentrations, and our knowledge of the Heisenberg limit is quite secure, an obvious small parameter is the number of holes. Another small parameter is the inverse spin size $1/S$, which controls the semiclassical expansion.

In this chapter, we extend the semiclassical analysis from the Heisenberg model to the $t - J$ model for a small number of holes. We begin by generalizing (19.1) from spin half to arbitrary spin S. We derive a spin-hole coherent state path integral representation which can be expanded by the method of steepest descents using S as the large parameter. The classical solutions are found to be small ferromagnetic polarons in the physically accessible regime of t/J. The semiclassical quantization of small polarons leads to an effective $t' - J$ Hamiltonian for the low-energy spin and charge excitations (in the ordered phase). A scenario for high-temperature superconductivity in the spin-liquid phase is mentioned.

19.1   Schwinger Bosons and Slave Fermions

The projector P_s is a *nonholonomic constraint* on the electron states:

$$P_s|n_i\rangle = \begin{cases} |n_i\rangle & n_i \leq 1 \\ 0 & n_i = 2 \end{cases}. \tag{19.3}$$

Usually, it is easier to handle *holonomic constraints* that are equalities rather than inequalities. This can be achieved by introducing two commuting Schwinger bosons (a_i, b_i), and a *"slave fermion"* f_i, at each site i, which obey the standard algebra

$$\begin{aligned} \left[a_i, a_j^\dagger\right] &= \delta_{ij}, \\ \left[b_i, b_j^\dagger\right] &= \delta_{ij}, \\ \{f_i, f_j^\dagger\} &= \delta_{ij}. \end{aligned} \tag{19.4}$$

The electronic states in the singly occupied space are represented by

$$\{c_{i\uparrow}^\dagger|0\rangle, c_{i\downarrow}^\dagger|0\rangle, |0\rangle\} = \{a_i^\dagger|0\rangle, b_i^\dagger|0\rangle, f_i^\dagger|0\rangle\}, \tag{19.5}$$

where $|0\rangle$ is the vacuum of all Schwinger bosons and slave fermions. The projector P_s becomes a holonomic constraint on the Schwinger boson and

---

[1]See Theorem 4.1.

slave fermion product Fock space

$$P_s \left(a_i^\dagger a_i + b_i^\dagger b_i + f_i^\dagger f_i - 1 \right) = 0, \qquad \forall \, i. \tag{19.6}$$

We can represent the projected electron operators by

$$c_{i\uparrow} P_s \;\; \rightarrow \;\; f_i^\dagger a_i P_s \, ,$$
$$c_{i\downarrow} P_s \;\; \rightarrow \;\; f_i^\dagger b_i P_s. \tag{19.7}$$

Thus, substituting (19.7) into the normal-ordered $t - J$ model of (3.33), we obtain

$$\mathcal{H}^{t-J} \;\; = \;\; P_s \Bigg[t \sum_{\langle ij \rangle} \left(f_i^\dagger f_j \mathcal{F}_{ji} + f_j^\dagger f_i \mathcal{F}_{ij} \right)$$

$$- \frac{J}{4} \sum_{\langle i,j,k \rangle} \left(\delta_{ik} - f_k^\dagger f_i \right) \mathcal{A}_{ij}^\dagger \mathcal{A}_{kj} \left(1 - f_j^\dagger f_j \right) \Bigg] P_s, \tag{19.8}$$

where $\langle ij \rangle$ denotes a nearest neighbor bond, and $\langle i, j, k \rangle$ denotes a sequence of three neighboring sites. The bond operators are defined as in (16.2) and (16.10):

$$\mathcal{F}_{ij} \;\; = \;\; a_i^\dagger a_j + b_i^\dagger b_j \, ,$$
$$\mathcal{A}_{ij} \;\; = \;\; a_i b_j - b_i a_j \, . \tag{19.9}$$

In the half filled (undoped) case, $n_i^f = 0$, the Schwinger bosons represent pure spin half operators. The hopping term in (19.8) vanishes, and the second term reduces to the antiferromagnetic Heisenberg model (see (16.13)).

Using this representation, we can generalize the $t - J$ model from spin one half to any spin S by simply replacing the constraint (19.6) by

$$P_S \left(a_i^\dagger a_i + b_i^\dagger b_i + f_i^\dagger f_i - 2S \right) = 0, \qquad \forall \, i. \tag{19.10}$$

This constraint implies that in the presence of a hole at site i, the spin at that site gets reduced by $\frac{1}{2}$. Incidentally, the large S spin hole representation may be realized in transition metal ions.[2]

19.2 Spin-Hole Coherent States

We wish to construct a representation of the $t - J$ model of spin S (19.8) that obeys the constraint (19.10). Let us first recall the *spin coherent states*

---

[2]where several d electrons are coupled ferromagnetically by Hund's rule; see Section 2.1.

of Section 7.3:

$$|\hat{\Omega}\rangle_S \equiv e^{iS\chi}\,[(2S)!]^{-1/2}\,\left[u(\theta,\phi)a^\dagger + v(\theta,\phi)b^\dagger\right]^{2S}|0,0\rangle . \tag{19.11}$$

χ is an arbitrary phase, and $\hat{\Omega}$ is a unit vector at latitude θ and longitude ϕ, and

$$\begin{aligned} u &= \cos(\theta/2)e^{i\phi/2} , \\ v &= \sin(\theta/2)e^{-i\phi/2} . \end{aligned} \tag{19.12}$$

The spin coherent states provide a resolution of the identity in the subspace of spin S:

$$\frac{2S+1}{4\pi}\int d\hat{\Omega}\,\,|\hat{\Omega}\rangle\langle\hat{\Omega}| = I, \tag{19.13}$$

where $d\hat{\Omega} = d\phi\,d\cos\theta$. In Chapter 10, this identity has been used to construct the spin coherent states path integral. Here we extend this basis to the Hilbert space of (19.10) and define the *"spin-hole coherent states"*:

$$|\hat{\Omega},\xi\rangle_S \equiv |\hat{\Omega}\rangle_S\,|0\rangle_f + |\hat{\Omega}\rangle_{S-\frac{1}{2}}\,\xi f^\dagger|0\rangle_f , \tag{19.14}$$

where ξ is an anticommuting Grassmann variable.[3]

The overlap of two states, with $\hat{\Omega} \approx \hat{\Omega}'$, is

$$\begin{aligned} \langle\hat{\Omega},\xi|\hat{\Omega}',\xi'\rangle &= \langle\hat{\Omega}|\hat{\Omega}'\rangle_S\left[1 + \xi^*\xi'((\langle\hat{\Omega}|\hat{\Omega}'\rangle_{\frac{1}{2}})^{-1}\right] \\ &\simeq \exp\left(-iS\psi[\hat{\Omega},\hat{\Omega}'] + e^{i\frac{1}{2}\psi[\hat{\Omega},\hat{\Omega}']}\xi^*\xi'\right), \end{aligned} \tag{19.15}$$

where

$$\begin{aligned} \psi &= 2\arctan\left\{\tan\left(\frac{\phi-\phi'}{2}\right)\frac{\cos[\frac{1}{2}(\theta+\theta')]}{\cos[\frac{1}{2}(\theta-\theta')]}\right\} + \chi - \chi' \\ &\simeq -\mathbf{A}\cdot(\hat{\Omega}-\hat{\Omega}'), \end{aligned} \tag{19.16}$$

and $\mathbf{A}$ is the familiar monopole vector potential given by (10.17) or (10.18). It obeys

$$\nabla\times\mathbf{A}\cdot\hat{\Omega} = 1. \tag{19.17}$$

The freedom in choosing $\chi(\hat{\Omega})$ in (19.11) is equivalent to the gauge freedom of $\mathbf{A}$. Thus, by (19.15) a gauge transformation on $\mathbf{A}$ must be accompanied by a $U(1)$ phase transformation on the Grassmann variable and vice versa,

$$\begin{aligned} \chi(\hat{\Omega}) &\rightarrow \chi'(\hat{\Omega}) , \\ \mathbf{A} &\rightarrow \mathbf{A}' = \mathbf{A} + \nabla_{\hat{\Omega}}(\chi'-\chi) , \\ \xi &\rightarrow e^{i(\chi'-\chi)/2}\xi. \end{aligned} \tag{19.18}$$

---

[3]See Appendix C.

This demonstrates the close relation between a gauge covariant coupling in (19.15) and the local constraints (19.10).

The spin-hole coherent states provide a resolution of the identity in the constrained Hilbert space

$$\frac{2S}{4\pi} \int d\hat{\Omega} d\xi^* d\xi \exp\left[-\alpha_S \, \xi^* \xi\right] |\hat{\Omega}, \xi\rangle\langle\hat{\Omega}, \xi| = I, \qquad (19.19)$$

where we have used (19.14) and the Gaussian Grassmann integral (C.10). The Gaussian normalization factor is

$$\alpha_S = \frac{2S+1}{2S}. \qquad (19.20)$$

We can now construct a path integral representation of the imaginary time generating functional using the resolution of the identity between each discrete timestep $\epsilon = \beta/N_\epsilon$. Thus, we obtain[4]

$$Z = \lim_{N_\epsilon \to \infty} \int \mathcal{D}\hat{\Omega}\, \mathcal{D}^2\xi \exp\left[-\sum_{i,\tau} \alpha_S \xi_i^*(\tau)\xi_i(\tau)\right]$$

$$\times \prod_{\tau=\epsilon}^{\beta} \langle\hat{\Omega}(\tau), \xi(\tau)| \, 1 - \epsilon\mathcal{H} \, |\hat{\Omega}(\tau-\epsilon), \xi(\tau-\epsilon)\rangle. \qquad (19.21)$$

The continuum limit is given by

$$Z = \int \mathcal{D}\hat{\Omega}\, \mathcal{D}^2\xi \, \exp\left\{\int_0^\beta d\tau \sum_i \left[i(S - \frac{1}{2}\xi_i^*\xi_i)\mathbf{A}(\hat{\Omega}_i) \cdot \dot{\hat{\Omega}}_i - \xi_i^* \partial_\tau^{(\alpha)} \xi_i\right]\right.$$

$$\left. -H^{t-J}[\hat{\Omega}, \xi^*, \xi] + \mu \sum_i \xi_i^* \xi_i\right\}. \qquad (19.22)$$

We recall[5] that treating the time derivative as if $\hat{\Omega}(\tau)$ is differentiable is being careless about operator ordering, which leads to uncertainties in the ground state energy. For the fermions, we do not need to assume continuity of $\xi(\tau)$. The discrete time derivative is

$$\int d\tau (\xi_i^* \partial_\tau^{(\alpha)} \xi_i) \equiv \sum_\tau \xi_i^*(\tau)\left[\alpha_S \xi_i(\tau) - \xi_i(\tau - \epsilon)\right]. \qquad (19.23)$$

(Do not worry about $\alpha_S \neq 1$; it will soon be fixed.)

---

[4]See Chapter 10 and Appendix D.
[5]See discussions of the spin kinetic term in Chapters 10 and 11.

The Hamiltonian function is

$$H^{t-J} = \frac{\langle \hat{\Omega}, \xi | \mathcal{H}^{t-J} | \hat{\Omega}, \xi \rangle}{\langle \hat{\Omega}, \xi | \hat{\Omega}, \xi \rangle} = H^t + H^J ,$$

$$H^t = \frac{\bar{t}}{\sqrt{2}} \sum_{\langle ij \rangle} \sqrt{1 + \hat{\Omega}_i \cdot \hat{\Omega}_j} \left(e^{\frac{i}{2} \psi_{ij}^F} \xi_i^* \xi_j + e^{-\frac{i}{2} \psi_{ij}^F} \xi_j^* \xi_i \right) ,$$

$$H^J = -\frac{\bar{J}}{8} \sum_{\langle i,j,k \rangle} e^{\frac{i}{2}(\psi_{ij}^N - \psi_{jk}^N)} \sqrt{(1 - \hat{\Omega}_j \cdot \hat{\Omega}_k)(1 - \hat{\Omega}_i \cdot \hat{\Omega}_j)}$$

$$\times (1 - \xi_j^* \xi_j) \xi_i^* \xi_k, \qquad (19.24)$$

where the (gauge dependent) phases ψ^F, ψ^N were defined in (17.5) and (17.10).

In order to obtain a finite classical limit for H^{t-J} at large S, we absorb the S dependence of (19.24) into the *classical parameters*, which scale as

$$4JS^2 \rightarrow \bar{J},$$
$$2tS \rightarrow \bar{t}. \qquad (19.25)$$

(These definitions reduce to the original parameters J, t for $S = \frac{1}{2}$.)

$H^{J'}$ includes fourth powers of Grassmann variables. These terms are hard to integrate, and an approximation is needed. In a Fock basis, the magnitude of a quartic term is

$$\left| \langle f_i^\dagger f_k f_j^\dagger f_j \rangle_{Fock} \right| = |\rho_{ik}| \rho_{jj} - \delta_{ik} \rho_{ij} \rho_{ji}, \qquad (19.26)$$

where ρ is the hole density matrix. In our classical solutions, we shall find that the hole density is primarily located in ferromagnetic regions where $\sqrt{1 - \hat{\Omega}_i \cdot \hat{\Omega}_j} \approx 0$. By (19.24) these are the regions where the ρ_{ij} is particularly small. Thus it is safe to replace H^{t-J} by the simplified bilinear Hamiltonian[6]:

$$H^{t-J} \approx \sum_{ij} \hat{H}_{ij}^t \xi_i^* \xi_j + \bar{H}^J[\hat{\Omega}, \rho] ,$$

$$\hat{H}_{ij}^t = \frac{\bar{t}}{\sqrt{2}} \sqrt{1 + \hat{\Omega}_i \cdot \hat{\Omega}_j} \, e^{i\psi_{ij}^F} ,$$

$$\bar{H}^J = \frac{\bar{J}}{4} \sum_{\langle ij \rangle} (\hat{\Omega}_i \cdot \hat{\Omega}_j - 1)(1 - \rho_i)(1 - \rho_j).$$

$$(19.27)$$

---

[6]Hamiltonian (19.27) often appears in the literature as the $t - J$ model.

In the third line we replaced the sum over triads $\langle i, j, k \rangle$ by a sum over bonds $\langle ij \rangle$ and omitted nondiagonal hopping terms since they are small compared to the t-hopping terms.

Now we integrate out the Grassmann variables[7] and obtain a *pure spin path integral*:

$$Z = \int \mathcal{D}\hat{\Omega} \exp \left\{ \int_0^\beta d\tau \left[i \sum_i (S - \rho_i/2) \mathbf{A}(\hat{\Omega}_i) \cdot \dot{\hat{\Omega}}_i - \bar{H}^J[\hat{\Omega}] \right.\right.$$

$$\left.\left. - F^f[\hat{\Omega}, \mu'] \right] \right\} . \tag{19.28}$$

$F^f[\mu]$ is the holes' (retarded) free energy functional

$$F^f = -\beta^{-1} \text{Tr}_i \ln \left[\alpha_S I + T_\tau \exp \left[-\int_0^\beta d\tau \left(\hat{H}^t_{ij}[\hat{\Omega}(\tau)] - \mu \delta_{ij} \right) \right] \right]$$

$$= -\beta^{-1} \text{Tr}_i \ln \left[I + T_\tau \exp \left[-\int_0^\beta d\tau \left(\hat{H}^t_{ij}[\hat{\Omega}(\tau)] - \mu' \delta_{ij} \right) \right] \right]$$

$$- \frac{1}{\epsilon} \mathcal{N} \ln(\alpha_S). \tag{19.29}$$

In the second line we absorb an infinite constant into the chemical potential

$$\mu' = \mu - \frac{1}{\epsilon} \ln \alpha_S. \tag{19.30}$$

The chemical potential and ground state energy shifts are *irrelevant* for the correlation functions, and we shall henceforth ignore them and set $\alpha_S \to 1$ and $\mu' \to \mu$.

Equation (19.28) is a useful starting point for the semiclassical approximation. As we take $S \to \infty$ the coefficient of the kinetic term increases, which suppresses time dependent spin fluctuations. This allows us to consider $F^f[\hat{\Omega}]$ and the density $\rho[\hat{\Omega}]$ as adiabatic functions of $\hat{\Omega}(\tau)$. That is to say: the large S justifies both semiclassical and adiabatic approximations.

19.3   The Classical Theory: Small Polarons

In the strict classical limit, the path integral is dominated solely by time-independent configurations

$$\hat{\Omega}(\tau) \to \hat{\Omega}(0) = \hat{\Omega}(\beta). \tag{19.31}$$

---

[7]See Appendix D, Section D.2.

By (10.23) Z is proportional to a classical partition function:

$$Z \to Z' \int \prod_i d\hat{\Omega}_i \exp\left[-\beta\left(\bar{H}^J[\hat{\Omega}] + F^f[\hat{\Omega}]\right)\right]. \qquad (19.32)$$

At zero temperature, the classical ground state for a fixed number of holes (N^f) is

$$
\begin{aligned}
E^{cl}[\hat{\Omega}] &= \bar{H}^J[\hat{\Omega}] + \lim_{T\to 0} F^f[\hat{\Omega}], \\
E^{cl}_0 &= \min_{\hat{\Omega}} \left. E^{cl} \right|_{\hat{\Omega}^{cl}, N^f},
\end{aligned}
\qquad (19.33)
$$

We first examine the case of a single hole.

The minimization of E^{cl}_0 for one hole is conceptually straightforward. Physically, there are two competing interactions: F^f prefers the largest number of highly connected ferromagnetic bonds in order to minimize the hole's kinetic energy. $\bar{H}^J$ prefers, of course, antiferromagnetic correlations. Candidate classical configurations may be either:

1. weakly distorted (canted or spiralling) Néel configurations, where the hole density is spread over the whole lattice, or

2. localized defects, where the hole density and the spin distortion are concentrated about a particular lattice position.

A variational minimization[8] has found that for

$$1 < \bar{t}/\bar{J} < 4 \qquad (19.34)$$

the classical configuration is of the second kind: a *five site polaron*, which is depicted in Fig. 19.1. At large values of $\bar{t}/\bar{J} \gg 1$, the polaron's radius grows as[9]

$$R_c \sim (\bar{t}/\bar{J})^{1/4}. \qquad (19.35)$$

In the limit of $\bar{t}/\bar{J} = \infty$, the whole lattice turns into a fully polarized ferromagnet. Interestingly, this agrees with the exact result of Nagaoka's Theorem 4.1 for the infinite U Hubbard model. We also learn that to achieve Nagaoka's limit for one hole requires U/t to exceed $\mathcal{O}(N^2)$.

The five site polaron does not distort the antiferromagnetic correlations of the background configuration. (Unlike, e.g., a single ferromagnetic bond, which induces a dipole shape distortion at large distances.) The "sharpness" of the polaron boundary is explained as follows: Both hole and Heisenberg energies depend *quadratically* on distortions at the polaron boundary. Since the stronger energy wins, the boundary spins are perectly aligned with the

---

[8] See Auerbach and Larson.

[9] See Exercise 1.

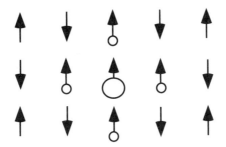

FIGURE 19.1. The five site polaron. Circles depict the hole density.

Néel directions. This implies that, beyond a short distance, two polarons have no classical interactions.

There is a gap of approximately $2\bar{t}$ in the fermion spectrum of the hole. The typical spin frequencies are

$$\omega \sim \mathcal{O}(\bar{J}S^{-1}) << \mathcal{O}(\bar{t}), \tag{19.36}$$

and therefore the adiabatic approximation for the holes' free energy is justified. This is similar to Born–Oppenheimer's approximation: The slow spins are like heavy nuclei, and the holes are like the fast electrons that determine an effective potential for the slow variables.

The spin-hole wave function of the five site polaron centered at site i is given by

$$|\hat{\Omega}^i, \rho\rangle = \sum_{j=i,\langle ij\rangle} \sqrt{\rho_j[\hat{\Omega}^i]}\, |\hat{\Omega}^i_j\rangle_{S-\frac{1}{2}} f_j^\dagger |0\rangle_{f_j} \prod_{j'\neq j}\left(|\hat{\Omega}^i_{j'}\rangle_S |0\rangle_{f_{j'}}\right), \tag{19.37}$$

where $\rho[\hat{\Omega}^i]$ is the ground state hole density for the polaron configuration $\hat{\Omega}^i$. For $S = \frac{1}{2}$, (19.37) can be created from the pure Néel state $\hat{\Omega}^N$ by

$$|\hat{\Omega}^i, \rho\rangle = p_{is_i}^\dagger |\hat{\Omega}^N\rangle, \tag{19.38}$$

where $p^\dagger$ is Dagotto and Schrieffer's quasiparticle operator

$$p_{is_i}^\dagger \equiv \left(\sqrt{\rho_i[\hat{\Omega}^i]} + \sum_{\langle ij\rangle,s'} \sqrt{\rho_j[\hat{\Omega}^i]}\, c_{is'}^\dagger c_{js'}\right) c_{is}. \tag{19.39}$$

$s_i = \pm\frac{1}{2}$ is the spin direction of $\hat{\Omega}_i^N$.

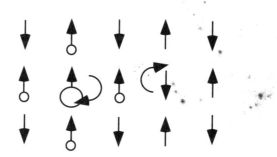

FIGURE 19.2. A polaron's tunneling path.

19.4   Polaron Dynamics and Spin Tunneling

The five site polaron breaks the spin rotational symmetry and lattice translational symmetry of the Hamiltonian. At finite S, even at zero temperature, symmetry can be restored by quantum fluctuations of the spins. The symmetry restoring fluctuations are classified into two distinct effects:

1. Spin waves, which tend to restore the spin symmetry. They contribute at order $\mathcal{O}(1/S)$.

2. Spin tunneling effects, which move the polaron's center and restore lattice translational symmetry. These are nonperturbative effects of order $\mathcal{O}(e^{-S(\cdot)})$.

Spin wave dynamics can be treated by the methods of Chapter 11 or 12 for the ordered and disordered phases, respectively.

The polaron's translational symmetry is discrete, and any motion of the spins involves climbing over an energy barrier. However, lattice symmetry can be restored by tunneling events where the central spin of the polaron rotates into an antiferromagnetic correlation with its neighbors, while another spin rotates and forms a five site ferromagnetic polaron elsewhere (see Fig. 19.2). The semiclassical tunneling matrix element for polaron motion from site i to k is[10]

$$\Gamma_{ik} \ \propto \ \langle \hat{\Omega}^i | G(E_0^{cl}) | \hat{\Omega}^k \rangle$$

---

[10]The correct multidimensional tunneling matrix element involves restricted wave functions and flux operators (see Auerbach and Kivelson). These will produce the correct prefactor Γ_0, but (19.40) can be used for the leading-order behavior.

$$= \int_{\hat{\Omega}^i}^{\hat{\Omega}^k} \mathcal{D}\hat{\Omega}(t) \exp\left\{ i \int_0^\infty dt \left[\sum_i (S - \rho_i/2)(1 - \cos\theta_i)\dot{\phi}_i \right.\right.$$

$$\left.\left. -(\bar{H}^J + E^f - E_0^{cl}) \right]\right\}, \qquad (19.40)$$

where $G(E)$ is the energy dependent Green's function defined in (10.38). By rotational symmetry, $H[\hat{\Omega}]$ is invariant under a global shift in all azimuthal angles

$$\phi_i = \bar{\phi} + \phi_i'. \qquad (19.41)$$

We integrate the kinetic term by parts and obtain

$$\int \mathcal{D}\bar{\phi}(t) \exp\left[-i \int dt \, \bar{\phi}(t)\partial_t \left(\sum_i [S - \rho_i/2(t)]\,[1 - \cos\theta_i(t)] \right) \right]$$

$$\propto \delta\left(M^z[\hat{\Omega}^i] - M^z[\hat{\Omega}^k] \right), \qquad (19.42)$$

where

$$M^z[\hat{\Omega}(t)] = \sum_j [S - \rho_j(t)/2]\,[1 - \cos\theta_j(t)] = M^z[\hat{\Omega}(0)]. \qquad (19.43)$$

That is to say, the z magnetization is conserved for any matrix element of $G(E)$. Of course, this is an exact consequence of the Hamiltonian's rotational invariance.

An important conclusion is that in the long range ordered phase, five site polarons cannot tunnel between sublattices A and B, since that would change the total magnetization. In other words, *there is a selection rule that forbids intersublattice hopping of polarons.* This strong result is somewhat surprising, since the original $t - J$ Hamiltonian (19.27) had large intersublattice hopping terms. Physically, the original f holes are strongly bound to a local spin defect. This restricts the charge mobility at low energies to intrasublattice hopping. We note, however, that polarons can tunnel between two remote sites on opposite sublattices if their local envirnoments have reversed Néel fields. (For example, the center and the tail of a Skyrmion configuration.)

The tunneling path is a saddle point of (19.40), which connects $\hat{\Omega}^i \to \hat{\Omega}^k$ and obeys the variational equations of motion

$$\frac{\delta}{\delta\hat{\Omega}_i(\tau)} \int_0^\infty d\tau' \left[(S - \rho_i/2)(1 - \cos\theta_i)\dot{\phi}_i - E^{cl}[\hat{\Omega}] \right] = 0. \qquad (19.44)$$

(We have fixed the gauge choice of **A** to carry out the tunneling rate calculation.)

In order to determine the tunneling path, we can make use of at least two conservation rules:

1. Conservation of energy along the classical path (see (10.37)),

$$E^{cl}[\hat{\Omega}^{tun}(\tau)] = E_0^{cl}. \tag{19.45}$$

2. Conservation of total magnetization (19.43).

Since $\hat{\Omega}^{tun}$ traverses a classically forbidden region, a solution can only be found by complexifying its coordinates.[11] We assume that the spins are ordered in the $\pm\hat{y}$ direction and parametrize the tunneling path by $\{\hat{\Omega}_i^{tun}(\varphi)\}$, where φ is the azimuthal rotation angle of spin i.

Now we complexify the ϕ variables of the tunneling path:

$$\phi(\varphi) \to \phi^{re}(\varphi) + i\phi^{im}(\varphi). \tag{19.46}$$

Using (19.45), the leading-order steepest descents approximation to (19.40) yields

$$\begin{aligned}
\Gamma_{ik} &\approx -\sum_\alpha \Gamma_0^\alpha \exp\left(-W_{ik}^\alpha + i\Upsilon_{ik}^\alpha\right), \\
W_{ik} &= \int d\varphi \sum_j \frac{d\phi_j^{im}}{d\varphi}(S - \rho_j/2)(1 - \cos\theta_j), \\
\Upsilon_{ik}^\alpha &= \int d\varphi \sum_j \frac{d\phi_j^{re}}{d\varphi}(S - \rho_j/2)(1 - \cos\theta_j),
\end{aligned} \tag{19.47}$$

where α labels the tunneling paths. Γ_0^α are positive prefactors, and the overall negative sign is general to tunneling matrix elements between degenerate minima in a time reversal invariant Hamiltonian.

From expressions (19.47) we deduce some qualitative information. First, we note that $|\Gamma_{ik}|$ decreases exponentially with S. Second, we see that the *sign* of Γ is determined by Υ, which is a Berry phase. Now we shall determine Υ for the tunneling of the five site polaron.

By symmetry under $\phi_i \to -\phi_i$, there are *two tunneling paths* $\hat{\Omega}^{tun,+}$ and $\hat{\Omega}^{tun,-}$, which involve clockwise and counterclockwise rotations in the xy plane:

$$\phi_i^{re,+}(\varphi) = -\phi_i^{re,-}(\varphi). \tag{19.48}$$

Thus, the two paths have the same values of W_{ik} but opposite phases Υ. By summing over both paths

$$\Gamma_{ik} \approx -\Gamma_0 \exp\left[-\int_0^\pi d\varphi \sum_i \left(S - \frac{\rho_i}{2}\right)(1 - \cos\theta_i)\frac{d\phi_i^{im}}{d\varphi}\right]$$

---

[11]The spin path integral is analogous to a *phase-space path integral*, where the conjugate variables are $\cos\theta, \phi$, see (11.7) and (11.5). Allowing ϕ to become imaginary is equivalent to making the momentum imaginary, i.e., continuing time to the imaginary axis in Feynman's path integral.

$$\times \sum_{\pm} \exp\left[i \int_0^{\pm\pi} d\varphi \sum_i \left(S - \frac{\rho_i}{2}\right)(1 - \cos\theta_i) \frac{d\phi_i^{re}}{d\varphi}\right]$$

$$= -2e^{i2\pi S} |\Gamma_{ik}| ,$$

(19.49)

where

$$|\Gamma_{ik}| = -\Gamma_0 \cos\left(\int_0^{\pm\pi} d\varphi \sum_i \frac{\rho_i}{2} \frac{d\phi_i^{re}}{d\varphi}\right)$$

$$\times \exp\left[-\int_0^{\pi} d\varphi \sum_i \left(S - \frac{\rho_i}{2}\right)(1 - \cos\theta_i) \frac{d\phi_i^{im}}{d\varphi}\right].$$

(19.50)

We use that $\sum_i \rho_i = 1$ to establish that the factor $\cos(\int d\varphi \ldots)$ is positive. The phase $2\pi S$ comes from the total rotations ϕ_i and ϕ_k. Other spins retrace their paths. Their Berry phases either vanish, or the contribution cancels by interference with a symmetric counterpart.

In (19.49) we discover that for half-odd integer spins, *the sign of the polaron hopping is positive*. This affects the ground state momentum as follows.

Assume a perfect Néel background. The band structure of the polaron hopping is

$$\epsilon_{\mathbf{k}} = \sum_j \Gamma_{ij} \exp[i\mathbf{k}(\mathbf{x}_i - \mathbf{x}_j)]$$

$$\approx 2\Gamma_{(2,0)}[\cos(2k_x) + \cos(2k_y)]$$
$$+2\Gamma_{(1,1)}[\cos(k_x + k_y) + \cos(k_x - k_y)], \quad (19.51)$$

where $(1,1), (2,0)$ denote first and second nearest neighbor sites on the same sublattice (see Fig. 19.3). We neglect further range hoppings which have a smaller tunneling rate. By (19.49) we see that for integer spins (and $\Gamma < 0$) $\epsilon_{\mathbf{k}}$ is minimized at $\mathbf{k} = 0$. For half-odd integer spins, the ground state momentum is located somewhere on the line

$$\mathbf{k}_0 \in \{ \mathbf{k} : \cos k_x + \cos k_y = 0 \}. \quad (19.52)$$

The sign of Γ_{ij} matters because intrasublattice hopping involves triangles, as seen, e.g., in Fig. 19.3. Therefore, its negative sign cannot be "gauged away" by redefining the wave function or choosing a different gauge convention in (19.44).

The semiclassical theory explains numerical studies of the spin half, $t - J$ model on small clusters.[12] First, the ground state momentum is found at

---

[12]See Dagotto and Schrieffer.

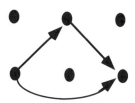

FIGURE 19.3. Dominant polaron hopping paths.

$(0, \pi)$ or $(\pi/2, \pi/2)$ (and their symmetric reflections), which agrees with (19.52). Second, the quasiparticle weight factor

$$Z_h = \left| \langle \Psi_0^{t-J} | p_{\mathbf{k}_0 s}^\dagger | \Psi_0^{Heis} \rangle \right|^2 / \langle \Psi_0^{t-J} | p_{\mathbf{k}_0 s} p_{\mathbf{k}_0 s}^\dagger | \Psi_0^{Heis} \rangle \qquad (19.53)$$

can be measured on small lattices. Here

$$p_{\mathbf{k}s}^\dagger = \sum_i e^{-i\mathbf{k}\cdot\mathbf{x}_i} p_{is}^\dagger \qquad (19.54)$$

and $p_{is}^\dagger$ is given by (19.39). $\Psi_0^{QHM}, \Psi_0^{t-J}$ are exact ground states of the Heisenberg model and the $t - J$ model with one hole, respectively. In the large S limit, we expect $Z_h \to 1$. Dagotto and Schrieffer numerically computed Z_h for the $t - J$ model on the 4×4 square lattice. For the parameter regime (19.34), they found

$$0.7 < Z_h < 0.94. \qquad (19.55)$$

This is a success of the semiclassical approximation for the $S = \frac{1}{2}$, $t - J$ model. The difference $1 - Z_h$ measures the quantum corrections to the short-range correlations of the five site polaron.

19.5 The $t' - J$ Model

For more than one hole, the classical ground states are obtained by filling the N^f lowest states of $\bar{H}^t$ and minimizing E^{cl} of (19.33).

We have found in Section 19.3 that the single polaron does not induce long-range distortions of the classical Néel state. Also, since the hole density is localized near the ferromagnetic bonds, the classical interactions between polarons are of *short range*. One must note, however, that quantum fluctuations (i.e., spin waves) will introduce longer-range interactions at higher orders in $1/S$.

For a finite density of holes, i.e., $\lim_{\mathcal{N} \to \infty} N^f/\mathcal{N} > 0$, one must consider the possibility of phase separation. The system may reduce its energy by crowding the holes into domains that share the distortions of the Néel background.[13]

Bearing in mind that the original Hubbard model neglects intersite repulsive Coulomb interactions, the short-range forces obtained by the classical approximation and the phase separation instability may be suppressed in the real copper oxide layers.

With the information at hand, however, it is possible to construct an effective low-energy Hamiltonian for the polaron quasiparticles. We substitute $|\hat{\Omega}^N\rangle$ in (19.39) by a slowly varying antiferromagnetically correlated configuration, and compute the polaron hopping rate as

$$\Gamma_{ik} \approx \delta_{s_i, s_k} \langle \hat{\Omega}^i | \bar{\Omega}^i \rangle \, \bar{\Gamma}_{ik} \, \langle \bar{\Omega}^k | \hat{\Omega}^k \rangle \,, \tag{19.56}$$

where $\bar{\Gamma}_{ik}$ is the unperturbed tunneling rate given by (19.49). $\bar{\Omega}^i$ and $\bar{\Omega}^k$ denote translated five site polarons centered at i and k, whose spins are precisely parallel to a common direction. This direction maximizes the overlaps with $\hat{\Omega}^i$ and $\hat{\Omega}^k$, respectively. The phases of the two overlap factors in (19.56) cancel for every spin except where the density of holes is different between $\hat{\Omega}^k$ and $\hat{\Omega}^i$. Using (19.37), one obtains

$$
\begin{aligned}
\langle \hat{\Omega}^i | \bar{\Omega}^i \rangle \langle \bar{\Omega}^k | \hat{\Omega}^k \rangle &\approx \exp \left[\frac{i}{2} \sum_j (2S - \rho_j^i) \cos \bar{\theta}_j^i (\phi_j^i - \bar{\phi}_j^i) \right] \\
&\quad \times \exp \left[-\frac{i}{2} \sum_j (2S - \rho_j^k) \cos \bar{\theta}_j^k (\phi_j^k - \bar{\phi}_j^k) \right] \\
&= \exp \left[-\frac{i}{2} \cos \bar{\theta} (\phi_i^i - \phi_k^k) \right],
\end{aligned}
\tag{19.57}
$$

where we have used that the total hole density in a polaron is unity. The lattice definition of the Néel gauge field A^N was given in (17.11). Here we find that polaron hopping is coupled to the Néel gauge field in a gauge covariant manner[14]:

$$\Gamma_{ik} = \bar{\Gamma}_{ik} \exp \left(i\eta_i \sum_\mu A^N \cdot \mathbf{x}_{ik} \right), \tag{19.58}$$

---

[13] See Emery, Kivelson, and Lin.
[14] See Shankar.

where $\eta_i = \text{sign}(s_i)$ is the effective "charge" of the polaron, which depends on its sublattice index.

The polaron operators $p_i^\dagger$ of (19.39) anticommute at distances greater than two lattice constants. Thus, the low-energy properties of a *dilute* system of polarons in the long-range ordered phase are described by

$$Z \propto \int \mathcal{D}\hat{\Omega} \mathcal{D}^2 p \, \exp\left(\int_0^\beta i\sum_i 2S\mathbf{A} \cdot \dot{\hat{\Omega}}_i - \frac{J}{4}\sum_{\langle ij \rangle} \hat{\Omega}_i \cdot \hat{\Omega}_j - L^p\right), \quad (19.59)$$

where L^p is a Lagrangian of spinless Grassmann variables p_i,

$$L^p = \sum_i p_i^* \left(\partial_\tau + i\eta_i A_0^N + \mu\right) p_i + \sum_{\langle i,j,k \rangle} \Gamma_{ik} e^{i\eta_i \mathbf{A}^N \cdot \mathbf{x}_{ik}} \, p_i^\dagger p_k$$

$$+ \sum_{ij} U_{ij} \, p_i^* p_i \, p_j^* p_j. \quad (19.60)$$

U_{ij} are the two-polaron effective interactions. The gauge invariance of L^p reflects the local constraints on the spin and charges as shown in (19.18). We can approximate the polarons and spin degrees of freedom by effective slave fermion and Schwinger boson operators that obey (19.10). A minimal Hamiltonian that corresponds to (19.60) is the $t' - J$ *model*,

$$\mathcal{H}^{t'-J,s} = P_s \left[\bar{J}\sum_{\langle ij \rangle} \mathbf{S}_i \cdot \mathbf{S}_j + \sum_{\langle i,k \rangle} \Gamma_{ik} p_i^\dagger p_k \left(a_i^\dagger a_k + b_i^\dagger b_k\right) \right] P_s, \quad (19.61)$$

or for spin half:

$$\mathcal{H}^{t'-J} = P_s \left(\bar{J}\sum_{\langle ij \rangle} \mathbf{S}_i \cdot \mathbf{S}_j - \sum_{\langle i,k \rangle,s} \Gamma_{ik} c_{is}^\dagger c_{ks} \right) P_s. \quad (19.62)$$

c_{is} are electron operators. The $t' - J$ model differs from the initial $t - J$ model by the absence of intersublattice hopping. The t' hopping has weaker local interaction with the spins, since the hole does not leave a trail of reversed spins when hopping on the same sublattice. This makes the $t' - J$ model more amenable to a weak coupling treatment of the hopping term, at least in the antiferromagnetic phase. Also, as noted in the following, it is possible to appeal to continuum approximations to describe the low-energy dynamics at low hole concentrations.

19.5.1   SUPERCONDUCTIVITY?

Following Haldane's mapping,[15] the spin interactions are described by the (2+1)-dimensional nonlinear sigma model with concentration dependent

---

[15]See Chapter 12.

stiffness constants and spin wave velocity. One expects that above some critical density, long-range spin order disappears but antiferromagnetic correlations survive at short length scales. As shown in (19.60), the low-energy spin fluctuations are coupled to the polarons via the Néel gauge field. An effective field theory for the coupled system has been proposed by Wiegmann, Wen, Shankar, and Lee (WWSL):

$$\mathcal{L}^{WWSL} = \sum_{\alpha=A,B} L^{p,\alpha}[-i\partial_\mu + \eta_\alpha A_\mu^N] + \frac{1}{m^2} \sum_{\mu\nu} (F_{\mu\nu})^2, \qquad (19.63)$$

where $L^{p,\alpha}$ is the long-wavelength Lagrangian of polarons on sublattice α, and the chiral spin fluctuations are described by an "electromagnetic field" derived from the Néel gauge field,

$$F_{\nu\mu} = \partial_\mu A_\nu^N - \partial_\nu A_\mu^N. \qquad (19.64)$$

m is the effective coupling constant of the Néel gauge field, which is roughly proportional to the inverse spin correlation length.[16]

WWSL have proposed that the ground state of (19.63) may be a high-temperature superconductor. The basic argument is that the polarons on different sublattices are oppositely charged with respect to the Néel gauge field. This gives rise to an electromagnetic-like attraction between polarons. It is a kinematical pairing mechanism which is not sensitive to the short-range interactions of the effective model. In addition, the scenario is valid in the magnetically disordered phase. This is particularly pleasing since, experimentally, antiferromagnetism does not seem to coexist with superconductivity in the cuprate superconductors.

19.6 Exercises

1. Assume a large circular region of radius R_c, where all spins are ferromagnetically correlated. Calculate the classical ground state energy E^{cl} for large $\bar{t}/\bar{J}$. Minimize the energy with respect to R_c to prove (19.35).

Bibliography

The Schwinger boson mean field theory for the $t - J$ model was worked out by

- C. Jayaprakash, H.R. Krishnamurthy, and S. Sarker, Phys. Rev. B **40**, 2610 (1989); C.L. Kane, P.A. Lee, T.K. Ng, B. Chakraborty, and N. Read, Phys. Rev. B **41**, 2653 (1990).

---

[16]See A.M. Polyakov, *Gauge Fields and Strings* (Harwood, 1987), Section 8.3.

The calculations of polaron size and interactions in the classical limit are found in

- A. Auerbach and B.E. Larson, Phys. Rev. Lett. **66**, 2262 (1991).

The gauge coupling of a hole in the fluctuating antiferromagnetic background was first derived in the continuum formulation by

- R. Shankar, Nucl. Phys. B **330**, 433 (1990).

Evaluation of multidimensional tunneling rates using restricted Green's functions (*"Path Decomposition Expansion"*) is reviewed in

- A. Auerbach and S. Kivelson, Nucl. Phys. Rev. B **257**, 799 (1985).

The definition of the polaron operator and the numerical evaluation of its weight for the $t - J$ model on small lattices is in

- E. Dagotto and J.R. Schrieffer, Phys. Rev. B **43**, 8705 (1991).

Variational arguments for phase separation in the $t - J$ model are given by

- V.J. Emery, S.A. Kivelson, and H.Q. Lin, Phys. Rev. Lett. **64**, 475 (1990).

Theories of high-temperature superconductivity in the $t' - J$ model were proposed by

- P.B. Wiegmann, Phys. Rev. Lett. **60**, 821 (1988);
- P.A. Lee, Phys. Rev. Lett. **63**, 690 (1989);
- X–G. Wen, Phys. Rev. B **39**, 7223 (1989).

Other approaches to the problem of holes in the two-dimensional antiferromagnet were developed by

- S. Trugman, Phys. Rev. B **41**, 892 (1990);
- B. Schraiman and E. Siggia, Phys. Rev. Lett. **62**, 156 (1989).

Part IV

Mathematical Appendices

$$
\begin{array}{r}
5\overset{\scriptstyle 1}{3}\,|\,\overset{\scriptstyle 1}{9} \\
+\,7283 \\
\hline
12{,}602
\end{array}
$$

$$
\begin{array}{r}
\times\;\overset{\scriptstyle 2\;+}{\overset{\scriptstyle .}{8}\!7}3 \\
24 \\
\hline
3{,}492 \\
+\,_1\;\,460 \\
\;1\;7{,} \\
\hline
20{,}952
\end{array}
$$

Calculation by Carmel A.

Appendix A

Second Quantization

Here we review some standard definitions and basic relations of second quantization, which are used throughout the text.

A.1 Fock States

One is given an orthonormal single-particle basis, $\{|\phi_i\rangle\}_{i=1}^{\mathcal{N}}$:

$$\langle \phi_i | \phi_j \rangle = \delta_{ij}. \tag{A.1}$$

The *creation operator* of state i is $a_i^\dagger$ and its Hermitian conjugate is the *annihilation operator* a_i. Both are defined with respect to the *vacuum state* $|0\rangle_i$, such that

$$|\phi_i\rangle = a_i^\dagger |0\rangle_i , \qquad a_i |0\rangle_i = 0 . \tag{A.2}$$

The number operator is defined as

$$n_i \equiv a_i^\dagger a_i. \tag{A.3}$$

Using the commutation rule

$$\left[a_i^\dagger a_i, (a_i^\dagger)^n \right] = n(a_i^\dagger)^n, \tag{A.4}$$

one verifies that repeated application of $a_i^\dagger$ (a_i) increases (decreases) the number of particles in the state ϕ_i. Fock states, labelled by *occupation numbers* n_i, are defined by

$$|n_1, n_2, \ldots\rangle = \prod_i \frac{\left(a_i^\dagger \right)^{n_i}}{\sqrt{n_i!}} |0\rangle_i . \tag{A.5}$$

For many bosons (or fermions), the symmetry (or antisymmetry) of the Fock states is enforced by commutation (or anticommutation) relations

$$\left[a_i, a_j^\dagger \right]_\eta = a_i a_j^\dagger - \eta a_j^\dagger a_i = \delta_{ij} , \tag{A.6}$$

$$\left[a_i, a_j \right]_\eta = 0 , \tag{A.7}$$

where

$$\eta = \begin{cases} 1 & \text{bosons} \\ -1 & \text{fermions} \end{cases} \tag{A.8}$$

The occupation numbers are restricted by the total number constraint

$$\sum_i n_i = N. \tag{A.9}$$

The commutation (or anticommutation) relations define an *algebra* of second quantized operators. This algebra is invariant under *canonical* transformations of the second quantized operators. Here we shall consider only *normal* transformations.[1] Consider a unitary transformation U on the single-particle basis

$$a_\alpha^\dagger = \sum_i U_{\alpha i} a_i^\dagger , \qquad U^\dagger U = I. \tag{A.10}$$

It is easy to verify that $\{a_\alpha^\dagger\}$ and $\{a_\alpha\}$ satisfy the canonical algebra (A.6) and (A.7).

A.2   Normal Bilinear Operators

Bilinear operators are given by

$$\hat{A} = \sum_{ij} a_i^\dagger A_{ij} a_j \equiv \mathbf{a}^\dagger \cdot A \cdot \mathbf{a}, \tag{A.11}$$

where A is a Hermitian matrix in the basis set ϕ_i, i.e., a single-particle operator.

The commutation relations of *bilinear* operators of bosons or fermions with *linear* operators are particularly simple. The vector $\mathbf{v}$ defines a linear operator $\hat{v}^\dagger$ as

$$\hat{v}^\dagger = \sum_i v_i a_i^\dagger = \mathbf{v} \cdot \mathbf{a}^\dagger . \tag{A.12}$$

By commuting (A.11) with (A.12) we find that

$$\left[\hat{A}, \hat{v}^\dagger\right] = (A\mathbf{v}) \cdot \mathbf{a}^\dagger. \tag{A.13}$$

If $\mathbf{v}$ is an eigenvector of A with eigenvalue v, then $\hat{v}^\dagger$ is an *eigenoperator* of $[A, \cdot]$ with eigenvalue v. This can be used to transform $\hat{v}^\dagger$ under the rotation:

$$\begin{aligned} e^{i\theta\hat{A}} \hat{v}^\dagger e^{-i\theta\hat{A}} &= \hat{v}^\dagger + i\theta[\hat{A}, \hat{v}^\dagger] + \frac{(i\theta)^2}{2}\left[\hat{A}, [\hat{A}, \hat{v}^\dagger]\right] + \cdots \\ &= e^{iv\theta}\hat{v}^\dagger. \end{aligned} \tag{A.14}$$

---

[1]For a treatise on Bogoliubov transformations, see the bibliography.

A unitary matrix U can be written in terms of a set of Hermitian generators A_α and parameters θ_α:

$$U_\theta = e^{i \sum_\alpha \theta_\alpha A_\alpha}. \tag{A.15}$$

A corresponding many-particle operator $\hat{U}$ is defined as

$$\hat{U}_\theta = e^{i \sum_\alpha \theta_\alpha \hat{A}_\alpha}. \tag{A.16}$$

Using (A.14), it is easy to show that $\hat{\mathbf{v}}^\dagger$ transforms under $\hat{U}$ in a simple manner:

$$\hat{U}_\theta \hat{\mathbf{v}}^\dagger \hat{U}_\theta^{-1} = (U_\theta \mathbf{v}) \cdot \mathbf{a}^\dagger. \tag{A.17}$$

The commutation relation between two bilinear operators is

$$\left[\hat{A}, \hat{B} \right] = \mathbf{a}^\dagger [A, B] \mathbf{a}. \tag{A.18}$$

(A.17) and (A.18) hold for both fermions and bosons.

EXAMPLE: SPIN ALGEBRA

The bilinear spin operators are

$$S^\alpha = \frac{1}{2} \sum_{ss'=1}^{2} a_s^\dagger \sigma_{ss'}^\alpha a_{s'}, \qquad \alpha = x, y, z, \tag{A.19}$$

where

$$(\sigma^x, \sigma^y, \sigma^z) = \left[\begin{pmatrix} 0 & 1 \\ 1 & 0 \end{pmatrix}, \begin{pmatrix} 0 & -i \\ i & 0 \end{pmatrix}, \begin{pmatrix} 1 & 0 \\ 0 & -1 \end{pmatrix} \right]. \tag{A.20}$$

By evaluating the commutation relations of the Pauli matrices, and by using (A.18), we can verify that S^α obeys the angular momentum relations

$$[S^\alpha, S^\beta] = i\epsilon^{\alpha\beta\gamma} S^\gamma, \tag{A.21}$$

where ϵ is the totally antisymmetric tensor.

A.3 Noninteracting Hamiltonians

Normal quadratic Hamiltonians are given by

$$\mathcal{H} = \sum_{ii'} a_i^\dagger H_{ii'} a_{i'} = \sum_k (\epsilon_k - \mu) a_k^\dagger a_k, \tag{A.22}$$

where H is Hermitian, ϵ_k, a_k are its energies and eigenoperators, respectively, and μ is the chemical potential. The partition Function is given by

$$
\begin{aligned}
Z &= \mathrm{Tr}\, e^{-\mathcal{H}/T} = \prod_k \sum_{n_k=0}^{n_{max}} \exp\left[-\frac{(\epsilon_k - \mu)n_k}{T}\right] \\
&= \prod_k (1 - \eta e^{-(\epsilon_k - \mu)/T})^{-\eta}.
\end{aligned}
\tag{A.23}
$$

In (A.23), n_{max} is infinity for bosons and one for fermions.

From (A.23), the free energies are

$$
F \equiv -T \ln Z = \eta T \sum_k \ln\left(1 - \eta e^{-(\epsilon_k - \mu)/T}\right).
\tag{A.24}
$$

The equilibrium occupation probabilities are given by

$$
\begin{aligned}
\langle n_k \rangle &= \frac{1}{Z} \mathrm{Tr}\left[a_k^\dagger a_k e^{-H/T}\right] = \frac{\partial F}{\partial \epsilon_k} \\
&= \left(e^{(\epsilon_k - \mu)/T} - \eta\right)^{-1},
\end{aligned}
\tag{A.25}
$$

which are the Bose and Fermi functions for $\eta = \pm 1$, respectively.

A.4   Exercises

1. Two Fock states of N bosons $\Phi_1 = |\{n_i\}\rangle$ and $\Phi_2 = |\{n_\alpha\}\rangle$ are defined with respect to the basis sets $\{\phi_i\}$ and $\{\phi_\alpha\}$, which are related by the unitary matrix U:

$$
\phi_\alpha = \sum_i U_{\alpha i} \phi_i.
\tag{A.26}
$$

Express the overlap $\langle \Phi_1 | \Phi_2 \rangle$ in terms of the matrix U.

2. Repeat Exercise 1, for the overlap of two fermion Fock states.

3. The operator of spin rotations is given by

$$
R = \exp[iS^z \phi] \exp[iS^y \theta] \exp[iS^z \chi].
\tag{A.27}
$$

Note that R is an operator defined in the fermion Fock (occupation number) space. Using (A.17), find the explicit expression for the SU(2) transformation matrix $U_{ss'}$,

$$
(c')_s = \left(RcR^{-1}\right)_s = \sum_{s'} U_{ss'}(\phi, \theta, \chi) c_{s'}.
\tag{A.28}
$$

4. Find the transformation under global spin rotation of the bilinear form:

$$
c_{1\uparrow}^\dagger c_{2\downarrow}^\dagger - c_{1\downarrow}^\dagger c_{2\uparrow}^\dagger,
\tag{A.29}
$$

where c_{1s} and c_{2s} represent electrons at two different orbitals.

5. Show that, for any global rotation given by (A.28), the paramagnetic Fermi gas state remains invariant:

$$\prod_{s,|k|<\pi/2} (c_{ks}^{\dagger})'|0\rangle = \prod_{s,|k|<\pi/2} c_{ks}^{\dagger}|0\rangle. \qquad (A.30)$$

Bibliography

Generalized Bogoliubov transformations are discussed in

- R. Balian and E. Brezin, Nuovo Cimento B **LXIV**, 37 (1969).

Appendix B

Linear Response and Generating Functionals

B.1 Spin Response Function

The spin response function describes the dynamics of the spins, in response to an arbitrary weak space-time dependent magnetic field (or *"source current"*) $j_i^\alpha(t)$, which is turned on at $t \geq 0$. The full source dependent Hamiltonian is

$$\mathcal{H}[j] = \mathcal{H}_0 - \sum_{\mathbf{q}} j_i^\alpha(t) S_i^\alpha, \tag{B.1}$$

where $\alpha = x, y, z$. At $t > 0$ all states in the Hilbert space evolve under the Schrödinger equation:

$$i\frac{\partial}{\partial t}|\psi(t)\rangle = \mathcal{H}[j]\,|\psi(t)\rangle. \tag{B.2}$$

The solution of (B.2) is given by the evolution operator

$$|\psi(t)\rangle = U[t, j]|\psi(0)\rangle\,,$$

$$U[t, j] = T_{t'} \exp\left(-i \int_0^t dt' \mathcal{H}[j(t')]\right), \tag{B.3}$$

where $\mathcal{H}_0(\mu)$ is the noninteracting Hamiltonian in the grand canonical formulation, and the *time ordered exponential* is defined as the limit of the discretized expression

$$T_{t'} \exp\left[-i \int_0^t dt' \mathcal{H}(t')\right]$$
$$= \lim_{\epsilon \to 0} \underbrace{[1 - i\mathcal{H}(t)\epsilon]\,[1 - i\mathcal{H}(t - 2\epsilon)\epsilon]\,\ldots[1 - i\mathcal{H}(\epsilon)\epsilon]}_{N_\epsilon}, \tag{B.4}$$

where $\epsilon = t/N_\epsilon$ is an infinitesimal *timestep*.

The expectation value of an operator S_i^α evolves under $\mathcal{H}[j)$ as

$$\langle S_i^\alpha(t)\rangle_j = Z^{-1} \operatorname{Tr}\left(\rho\, U^{-1}[t, j] S_i^\alpha U[t, j]\right)\,, \tag{B.5}$$

where $Z = \mathrm{Tr}\rho$ is the partition function and ρ is the density matrix. To linear order in the source current j, (B.5) yields

$$\langle S_i^\alpha(t)\rangle_j = \langle S_i^\alpha(0)\rangle + i \int_0^t dt' \sum_{i',\alpha'} j_{i'}^{\alpha'} \langle\left[S_i^\alpha(t), S_{i'}^{\alpha'}(t')\right]\rangle + \mathcal{O}(j^2), \quad \text{(B.6)}$$

where $\langle\cdot\rangle$ is defined with respect to $\mathcal{H}_0$. The kernel of the integral in (B.6) is the response function

$$R_{ii'}^{\alpha\alpha'}(t - t') = -i\theta(t - t') \langle[S_i^\alpha(t), S_{i'}^\alpha(t')]\rangle. \quad \text{(B.7)}$$

Due to the θ function, R is also called the "retarded" correlation function. Henceforth we shall drop the superscripts α, α'.

It is possible to calculate R using the eigenstates and eigenenergies of $\mathcal{H}_0$:

$$\mathcal{H}_0(\mu) |\alpha\rangle = E_\alpha|\alpha\rangle. \quad \text{(B.8)}$$

In the eigenstates representation, (B.7) is given by

$$R_{ii'}(t) = -i\theta(t)Z^{-1}\sum_{\alpha\beta} e^{-E_\alpha/T}\left(e^{i(E_\alpha - E_\beta)t}\langle\alpha|S_i^\alpha|\beta\rangle\langle\beta|S_{i'}^\alpha|\alpha\rangle\right.$$

$$\left. -e^{i(E_\beta - E_\alpha)t}\langle\alpha|S_{i'}^\alpha|\beta\rangle\langle\beta|S_i^\alpha|\alpha\rangle\right). \quad \text{(B.9)}$$

Translational invariance of the Hamiltonian yields $R_{ii'} = R(\mathbf{x}_i - \mathbf{x}_{i'})$, where $\mathbf{x}_i$ is the position of lattice site i. The space-time Fourier transform of R is

$$R(\mathbf{q}, \omega + i0^+) = \mathcal{N}^{-1}\sum_{ii'} \int_0^\infty dt \, e^{-i\mathbf{q}(\mathbf{x}_i - \mathbf{x}_{i'}) + i(\omega + i0^+)t} R(\mathbf{x}_i - \mathbf{x}_{i'}, t)$$

$$= (\mathcal{N}Z)^{-1}\sum_{\alpha,\beta} \langle\alpha|S_{\mathbf{q}}^\alpha|\beta\rangle\langle\beta|S_{-\mathbf{q}}^\alpha|\alpha\rangle \frac{e^{-E_\alpha/T} - e^{-E_\beta/T}}{E_\alpha - E_\beta + \omega + i0^+},$$

$$\text{(B.10)}$$

where

$$S_{\mathbf{q}}^\alpha = \sum_i e^{-i\mathbf{q}\mathbf{x}_i} S_i^\alpha. \quad \text{(B.11)}$$

The infinitesimal positive number 0^+ ensures convergence of the Fourier integral at large times. The real and imaginary parts of R are given by the identity

$$\frac{1}{x + i0^+} = \mathcal{P}(\frac{1}{x}) - i\pi\delta(x), \quad \text{(B.12)}$$

where $\mathcal{P}$ takes the principal part of its argument. Applying (B.12) to (B.10), one can prove the Kramers–Kronig relations

$$\text{Re } R(\mathbf{q}, \omega) = \mathcal{P} \int_{-\infty}^{\infty} \frac{d\omega'}{\pi} \frac{\text{Im } R(\mathbf{q}, \omega')}{\omega' - \omega},$$

$$\text{Im } R(\mathbf{q}, \omega) = -\mathcal{P} \int_{-\infty}^{\infty} \frac{d\omega'}{\pi} \frac{\text{Re } R(\mathbf{q}, \omega')}{\omega' - \omega}. \tag{B.13}$$

B.2   Fluctuations and Dissipation

The imaginary part of the spectral response function $\text{Im } R_{21}(\omega)$ is the dissipative part, which describes the damping of the external oscillations at frequency ω by exciting internal modes of the system. The correlation function between two spin fluctuations at equilibrium is

$$S_{i,i'}^{\alpha\alpha'}(t - t') = \langle S_i^{\alpha}(t) \cdot S_{i'}^{\alpha'}(t') \rangle. \tag{B.14}$$

Its Fourier transform is called the *dynamical structure factor*:

$$S^{\alpha\alpha'}(\mathbf{q}, \omega) = \mathcal{N}^{-1} \sum_{ii'} \int_{-\infty}^{\infty} dt \, e^{-i\mathbf{q}(\mathbf{X}_i - \mathbf{X}_{i'}) + i\omega t} S_{ii'}^{\alpha\alpha'}(t)$$

$$= \frac{2\pi}{ZN} \sum_{\alpha,\beta} e^{-E_\alpha/T} \langle \alpha | S_{\mathbf{q}}^{\alpha} | \beta \rangle \langle \beta | S_{-\mathbf{q}}^{\alpha'} | \alpha \rangle \, \delta(\omega + E_\alpha - E_\beta). \tag{B.15}$$

For Hermitian observables, the *structure factor* $S(\mathbf{q}, \omega)$ is real. It describes the spontaneous fluctuations at frequency ω and is measurable by scattering experiments. For example, a polarized inelastic neutron scattering cross section measures the electronic spin structure factor.

By comparing the explicit expressions (B.10) and (B.15), we obtain a simple relation between the dissipative response function and the structure factor (dropping the superscripts)

$$S(\mathbf{q}, \omega) = \frac{2}{1 - e^{-\omega/T}} \text{Im } R(\mathbf{q}, \omega). \tag{B.16}$$

This relation is called the *fluctuation-dissipation theorem*.

B.3   The Generating Functional

Theorists are interested in calculating response functions or structure factors from the microscopic model. A useful device which allows such calculations in interacting many-particle systems is the *imaginary time generating functional.* In the text, we encounter path integral representations of the generating functional, which are amenable to the semiclassical or large N approximations. Here we derive the response function $R(\mathbf{q}, \omega)$ as a second derivative of the generating functional.

The grand canonical generating functional is defined as

$$Z[j, \mu] = \operatorname{Tr} T_\tau \left(\exp \left[- \int_0^\beta d\tau \, \mathcal{H}[j(\tau)] \right] \right), \qquad (B.17)$$

where

$$\mathcal{H}[j(\tau)] = \mathcal{H}_0(\mu) - \sum_i j_i^\alpha(\tau) \, S_i^\alpha , \qquad \tau \in [0, \beta). \qquad (B.18)$$

The time ordering operator $T(\bullet)$, previously used in (B.4), orders the operators such that τ increases from right to left.

For a rotationally symmetric $\mathcal{H}_0$, $\langle S_i^\alpha \rangle = 0$. The imaginary correlation function is defined as

$$
\begin{aligned}
\tilde{R}_{i,i'}^{\alpha\alpha'}(\tau, \tau') &= \left. \frac{\delta^2 \ln Z}{\delta j_i^\alpha(\tau) \delta j_{i'}^{\alpha'}(\tau')} \right|_{j=0} \\
&= Z^{-1} \operatorname{Tr} \left\{ e^{-\beta \mathcal{H}_0} \, T_\tau [S_i^\alpha(\tau) S_{i'}^\alpha(\tau')] \right\}, \qquad (B.19)
\end{aligned}
$$

where

$$S_i^\alpha(\tau) \equiv e^{\mathcal{H}_0 \tau} S_i^\alpha e^{-\mathcal{H}_0 \tau}, \qquad (B.20)$$

and the time ordering is defined as

$$T_\tau [A(\tau)B(\tau')] = \begin{cases} A(\tau)B(\tau') & \tau \geq \tau' \\ B(\tau')A(\tau) & \tau < \tau' \end{cases}. \qquad (B.21)$$

From (B.19), using the cyclic property of the trace, we verify that

$$\tilde{R}(\tau, \tau') = \tilde{R}(\tau - \tau', 0) \equiv \tilde{R}(\tau - \tau'), \qquad (B.22)$$

and that $\tilde{R}(\tau)$ is periodic on the interval $[0, \beta)$:

$$\lim_{\tau \to \beta^-} \tilde{R}(\tau) = \lim_{\tau \to 0^-} \tilde{R}(\tau). \qquad (B.23)$$

For translationally invariant Hamiltonians in space and time, it is convenient to use the Fourier representation of $\tilde{R}_{ii'}(\tau)$ given by

$$\tilde{R}(\mathbf{q}, i\omega_n) = \mathcal{N}^{-1} \sum_{ii'} \int_0^\beta d\tau \, e^{-i\mathbf{q}(\mathbf{X}_i - \mathbf{X}_{i'}) - i\omega_n \tau} \tilde{R}_{ii'}(\tau), \qquad (B.24)$$

$$\omega_n = 2n\pi\beta^{-1}, \qquad n = 0, \pm 1, \pm 2, \ldots. \qquad (B.25)$$

The frequencies ω_n are the *Bose–Matsubara frequencies*. By inserting a complete set of eigenstates between the operators in (B.19) and performing the $d\tau$ integration in (B.24), we find that $R(\mathbf{q}, i\omega_n)$ is indeed the analytic continuation of the spectral response function given by (B.10):

$$\tilde{R}(\mathbf{q}, z)\Big|_{z \to \omega + i0^+} = \operatorname{Re} R(\mathbf{q}, \omega) + i \operatorname{Im} R(\mathbf{q}, \omega). \tag{B.26}$$

The functions on the right-hand side are the physically measurable quantities that describe the system's response and dissipation at frequency ω. These are related to the structure factor by the fluctuation-dissipation theorem (B.16). The left-hand side can be obtained from theory, using, e.g., the imaginary time generating functional.

Finally, we define the static susceptibility as the response of the magnetization at momentum $\mathbf{q}$ to an ordering field in the α direction, of wave vector $\mathbf{q}$:

$$\chi^{\alpha\alpha}(\mathbf{q}) = \frac{1}{2} \operatorname{Re} R^{\alpha\alpha}(\mathbf{q}, 0). \tag{B.27}$$

Bibliography

Some recommended textbooks for this material are:

- P.C. Martin, *Measurements and Correlation Functions* (Gordon and Breach, 1968);

- G. Rickayzen, *Green's Functions and Condensed Matter* (Academic Press, 1980).

See also the bibliographies of Chapters 1 and 2.

Appendix C

Bose and Fermi Coherent States

Coherent states form overcomplete bases, which are labelled by a continuous parameter or variable. The resolution of the identity in the coherent state basis is often useful for evaluating matrix elements and traces of operators. Here we shall define the states and present several useful identities which follow directly from the definitions.

C.1 Complex Integration

Bose coherent states are parametrized by complex vectors

$$\mathbf{z} = (z_1, z_2, \ldots, z_{\mathcal{N}}), \quad z_i = x_i + iy_i, \tag{C.1}$$

where x_i, y_i are real and $\mathcal{N}$ is the number of single-particle states. The complex integration is defined as

$$\int d^2 \mathbf{z} \equiv \int_{-\infty}^{\infty} \prod_i \left(\frac{dx_i \, dy_i}{\pi} \right). \tag{C.2}$$

A useful identity of complex integration over a single variable z is

$$\frac{1}{n!} \int d^2 \mathbf{z} (z^*)^n z^m \exp(-z^* z) = \delta_{n,m}. \tag{C.3}$$

Using this definition, it follows that for any complex matrix G, whose Hermitian part has only positive eigenvalues, the Gaussian integral is given by

$$\int d^2 \mathbf{z} \, e^{-\mathbf{z}^* G \mathbf{z} \, -\mathbf{z}_a^* \mathbf{z} - \mathbf{z}^* \mathbf{z}_b} = \det |G|^{-1} e^{\mathbf{z}_a^* G^{-1} \mathbf{z}_b}, \tag{C.4}$$

where $\mathbf{z}_a^*, \mathbf{z}_b$ are any complex vectors.

C.2 Grassmann Variables

Fermion coherent states are parametrized by conjugated pairs of Grassmann vectors

$$\mathbf{z} = (z_1, z_2, \ldots, z_{\mathcal{N}}), \quad \mathbf{z}^* = (z_1^*, z_2^*, \ldots, z_{\mathcal{N}}^*), \tag{C.5}$$

where $\mathcal{N}$ is the number of single-particle states. These variables are *operators* which can be added and multiplied. They commute with any scalar c, but anticommute with each other:

$$
\begin{aligned}
cz &= zc, \\
z_i z_j &= -z_j z_i, \\
z_i z_j^* &= -z_j^* z_i.
\end{aligned}
$$
$$\text{(C.6)}$$

(The conjugate of a Grassmann variable is treated as a different variable.) Equation (C.6) implies that

$$z_i^2 = 0. \tag{C.7}$$

Thus, any function of Grassmann variables can be expanded as a finite sum.

The *"Grassmann integration"* is a counting operation which acts on a product of Grassmann variables as follows:

$$\int dz_1 \prod_{i=1}^{\mathcal{N}} z_i^{n_i} \prod_{j=1}^{\mathcal{N}} (z_j^*)^{m_j} c = n_1 \prod_{i=2}^{\mathcal{N}} z_i^{n_i} \prod_{j=1}^{\mathcal{N}} (z_j^*)^{m_j} c, \tag{C.8}$$

where $n_i, m_j = 0, 1$, and c is any scalar number. For a general product of Grassmann's, we first anticommute the variable to be integrated to the left, and obtain an overall sign factor of the permutation. Since z_1 and z_1^* are treated as two different variables, we define the integration over $d^2 z_1$ as

$$\int d^2 z_1 O(\mathbf{z}^*, \mathbf{z}) \equiv \int dz_1^* \left[\int dz_1 O(\mathbf{z}^*, \mathbf{z}) \right], \tag{C.9}$$

where O is a sum of products. It follows from (C.8) that the contribution of each product vanishes unless it contains exactly one power of z_1 and one power of z_1^*. From the definitions above, it follows that for one variable and its complex conjugate, the Grassmann Gaussian integral is

$$\int d^2 z \, \exp\left(-z^* z\right) z^m (z^*)^n = \delta_{mn}, \quad n, m = 0, 1, \tag{C.10}$$

which is formally similar to the complex integral (C.3). The multidimensional Gaussian integral is

$$\int d^2 \mathbf{z} \, \exp\left(-\mathbf{z}^* G \mathbf{z} + \mathbf{z}_a^* \mathbf{z} + \mathbf{z}^* \mathbf{z}_b\right) = \det|G| \, \exp\left(\mathbf{z}_a^* G^{-1} \mathbf{z}_b\right), \tag{C.11}$$

where $\mathbf{z}_a^*, \mathbf{z}_b$ are Grassmann vectors. Equations (C.4) and (C.11) can be unified by the single expression

$$\int d^2 \mathbf{z} \, \exp\left(-\mathbf{z}^* G \mathbf{z} + \mathbf{z}_a^* \mathbf{z} + \mathbf{z}^* \mathbf{z}_b\right) = \det|G|^{-\eta} \, \exp\left(\mathbf{z}_a^* G^{-1} \mathbf{z}_b\right), \tag{C.12}$$

where

$$\eta = \begin{cases} 1 & \text{bosons} \\ -1 & \text{fermions} \end{cases}. \tag{C.13}$$

C.3   Coherent States

Coherent states are defined as

$$|\mathbf{z}\rangle = \exp\left(\mathbf{z}\mathbf{a}^\dagger\right)|0\rangle, \tag{C.14}$$

where $\mathbf{a}$ is a vector of Bose (Fermi) creation operators, and $\mathbf{z}$ is a vector of complex numbers (Grassmann variables). Coherent states are eigenstates of annihilation operators

$$a_i|\mathbf{z}\rangle = z_i|\mathbf{z}\rangle. \tag{C.15}$$

The overlap of two coherent states is

$$\langle\mathbf{z}|\mathbf{z}'\rangle = e^{\mathbf{z}^*\mathbf{z}'}. \tag{C.16}$$

Thus, the basis $\{|\mathbf{z}\rangle\}$ is not orthogonal but it spans Fock space, as can be seen by resolution of the identity given by the integral

$$\int d^2z\, e^{-\mathbf{z}^*\mathbf{z}}\,|\mathbf{z}\rangle\langle\mathbf{z}| = I, \tag{C.17}$$

which follows from (C.3) and (C.10).

Any operator that can be expanded in terms of creation and annihilation operators can be *normal ordered*, i.e., written as a sum of products where each product has the creation operators to the left of the annihilation operators. We therefore restrict our discussion to normal ordered operators of the form

$$\begin{aligned}
\mathcal{O}[\mathbf{a}^\dagger,\mathbf{a}] &= \sum_{\{n_i,m_i\}} O_{\{n_i,m_i\}}\left[(a_1^\dagger)^{n_1}\cdots(a_N^\dagger)^{n_N}\,a_1^{m_1}\cdots a_N^{m_N}\right], \\
&= :\mathcal{O}:, \tag{C.18}
\end{aligned}$$

where for fermions, $n_i, m_j \leq 1$. Using (C.15), the matrix element of any normal ordered operator between coherent states is

$$\langle\mathbf{z}|\mathcal{O}[\mathbf{a}^\dagger,\mathbf{a}]|\mathbf{z}'\rangle = O(\mathbf{z}^*,\mathbf{z}')e^{\mathbf{z}^*\mathbf{z}'}, \tag{C.19}$$

where

$$O(\mathbf{z}^*,\mathbf{z}') = \sum_{\{n_i,m_i\}} O_{\{n_i,m_i\}}\left[(z_1^*)^{n_1}\cdots(z_N^*)^{n_N}\,z_1^{m_1}\cdots z_N^{m_N}\right], \tag{C.20}$$

where for fermions, $n_i, m_j \leq 1$.

Finally, it is easy to prove, using (C.17), that the trace of any operator in Fock space is given by

$$\operatorname{Tr} A = \int d^2\mathbf{z}\, e^{-\mathbf{z}^*\mathbf{z}}\langle\mathbf{z}|\mathcal{O}|\eta\mathbf{z}\rangle. \tag{C.21}$$

Note the sign factor η, which arises from changing the order of Grassmann variables in the process of proving this equality.

C.4 Exercises

1. Prove the identities given in the text for complex integrals and Bose coherent states ($\eta = 1$): (C.3), (C.4), (C.16), (C.17), and (C.19).

2. Prove the identities for the Grassmann integrals and Fermi coherent states ($\eta = -1$): (C.10), (C.11), (C.16), (C.17), and (C.19).

3. Use the resolutions of the identity (C.17) to show that two operators having the same matrix elements between any two boson or fermion coherent states are identical.

4. *Normal ordering of exponentials.* You are given an exponential bilinear operator of one state (either Bose or Fermi statistics):

$$B = \exp\left[\lambda a^\dagger a\right]. \tag{C.22}$$

Show that both for the Bose and Fermi cases,

$$\langle z|B|z'\rangle = \exp\left[e^\lambda z^* z'\right]. \tag{C.23}$$

5. Use the previous problem and (C.19) to show that B can be written as

$$B =: \exp\left[(e^\lambda - 1)a^\dagger a\right] : , \tag{C.24}$$

where $: \mathcal{O} :$ is given by normal ordering each product of operators in $\mathcal{O}$.

6. Generalize (C.24) to many particles, and show that for a Hermitian matrix Λ:

$$\begin{aligned} B &= \exp\left[\mathbf{a}^\dagger \Lambda \mathbf{a}\right] \\ &= : \exp\left[\mathbf{a}^\dagger \left(e^\Lambda - I\right)\mathbf{a}\right] : . \end{aligned} \tag{C.25}$$

(*Hint: use the eigenbasis of Λ.*)

Bibliography

Some recommended references for this material are

- J.W. Negele and H. Orland, *Quantum Many Particle Systems* (Addison-Wesley, 1988), Chapter 1 ;

- L.S. Schulman, *Techniques and Applications of Path Integration* (Wiley, 1981), Chapter 27.

Appendix D

Coherent State Path Integrals

D.1 Constructing the Path Integral

The imaginary time generating functional (see (B.17)) is

$$Z[j,\mu] = \operatorname{Tr} T_\tau \left(\exp\left[-\int_0^\beta d\tau\, \mathcal{H}[j(\tau)] \right] \right), \tag{D.1}$$

where

$$\mathcal{H}[j(\tau)] = \mathcal{H}_0(\mu) - \sum_{i,\alpha} j_i^\alpha(\tau) a_{i,s}^\dagger \sigma_{ss'}^\alpha a_{i,s'}, \qquad \tau \in [0,\beta). \tag{D.2}$$

The source currents j are coupled to spin operators.[1] The trace formula (C.21) yields the representation

$$
\begin{aligned}
Z &= \operatorname{Tr} T_\tau \left(\exp\left[-\int_0^\beta d\tau\, \mathcal{H}[j(\tau)] \right] \right) \\
&= \int d^2 \mathbf{z}_0\, e^{-\mathbf{z}_0^* \mathbf{z}_0} \langle \mathbf{z}_0 | T \left(\exp\left[-\int_0^\beta d\tau\, H(\tau) \right] \right) |\eta \mathbf{z}_0 \rangle, \tag{D.3}
\end{aligned}
$$

where $\mathbf{z} = (z_1, \ldots, z_\mathcal{N})$ are the Bose (Fermi) coherent state variables defined on a single-particle basis set of $\mathcal{N}$ states. $\eta = 1$ ($\eta = -1$) for bosons (fermions).

The time ordered exponential is defined as

$$T_\tau \left(e^{-\int_0^\beta d\tau\, \mathcal{H}(\tau)} \right) = \lim_{\epsilon \to 0} T_n \prod_{n=1}^{N_\epsilon} [1 - \epsilon \mathcal{H}(n\epsilon)], \tag{D.4}$$

where $\epsilon = \beta/N_\epsilon$ is the timestep, and T_n orders the factors such that n decreases from left to right. Between each factor in (D.4) we can insert a resolution of the identity using Bose or Fermi coherent states (see (C.17)).

---

[1] Here we restrict ourselves to spin source terms. The formalism can be applied to source terms of any operators.

This introduces N_ϵ integration variables $\mathbf{z}_\tau$. The endpoints are identified (up to a factor of η), such that

$$\mathbf{z}_\beta = \eta \mathbf{z}_0. \tag{D.5}$$

The discrete generating functional is expressed as a multivariate complex or Grassmann integral:

$$Z(\epsilon) = \int \prod_{\tau=\epsilon}^{\beta} (d^2 \mathbf{z}_\tau) \exp\left[-\sum_{\tau=\epsilon}^{\beta} \mathbf{z}_\tau^*(\mathbf{z}_\tau - \mathbf{z}_{\tau-\epsilon}) - \sum_{\tau=\epsilon}^{\beta} H(\tau)\epsilon \right], \tag{D.6}$$

where we used (C.19) to define $H[\mathbf{z}^*, \mathbf{z}]$ as

$$\langle \mathbf{z}_\tau | \mathcal{H} | \mathbf{z}_{\tau-\epsilon} \rangle = H[\mathbf{z}_\tau^*, \mathbf{z}_{\tau-\epsilon}, j(\tau)]\, e^{\mathbf{z}_\tau^* \mathbf{z}_{\tau-\epsilon}}, \tag{D.7}$$

and $(1 - \epsilon H) = e^{-\epsilon H} + \mathcal{O}(\epsilon)$. The limit of $\epsilon \to 0$ is formally denoted by the path integral

$$Z = \lim_{\epsilon \to 0} Z(\epsilon)$$

$$\equiv \int \mathcal{D}^2 z(\tau) \exp\left(-\int_0^\beta d\tau \, \{ \mathbf{z}^* \partial_\tau \mathbf{z} + H[j(\tau)] \} \right). \tag{D.8}$$

The *"time derivative"* or *"kinetic term"* is defined in the discrete formulation

$$\partial_\tau \equiv (\delta_{\tau,\tau'} - \delta_{\tau-\epsilon,\tau'})/\epsilon. \tag{D.9}$$

The kinetic term partially suppresses the contributions of rapidly varying paths.

Caution: Since the integration variables at consecutive timesteps are totally independent, one cannot assume continuity or smoothness of the paths $\mathbf{z}(\tau)$!

D.2   Normal Bilinear Hamiltonians

An important subset of generating functionals involves normal bilinear Hamiltonians,

$$\int_0^\beta d\tau \, H(\tau) = \sum_{\tau i i'} z_{i\tau}^* h_{i,i'}(\tau) z_{i'\tau-\epsilon}\, \epsilon. \tag{D.10}$$

The generating functional $Z(\epsilon)$ for normal bilinear Hamiltonians is a multidimensional Gaussian integral, which by (C.12) yields

$$Z(\epsilon) = \int \prod_{\tau=0}^{\beta-\epsilon} (d^2 \mathbf{z}_\tau) \exp\left[-\epsilon \sum_{i\tau, i'\tau'} z_{i\tau}^* G_{i\tau, i'\tau'}^{-1} z_{i'\tau'} \right]$$

$$= (\det G\epsilon)^\eta$$

$$= \exp[\eta \mathrm{Tr} \ln(G\epsilon)], \tag{D.11}$$

where the Green's function is given by

$$G^{-1}\epsilon = (\hat{\partial}_\tau + \hat{h})\epsilon \ . \tag{D.12}$$

G can be written explicitly as a matrix of size $(N_\epsilon \mathcal{N})^2$, in the spatial and temporal representation:

$$G^{-1}\epsilon = \begin{pmatrix} I & 0 & 0 & \cdots & \eta(-I + \epsilon h(\beta)) \\ -I + \epsilon h(0) & I & 0 & \cdots & 0 \\ 0 & -I + \epsilon h(\epsilon) & I & 0 & \vdots \\ \vdots & \ddots & & \ddots & \ddots \\ 0 & \cdots & & -I + \epsilon h(\beta - \epsilon) & I \end{pmatrix} \tag{D.13}$$

By successively eliminating the last timestep blocks using the identity

$$\det \begin{pmatrix} A & B \\ C & D \end{pmatrix} = \det D \det \left(A - BD^{-1}C \right), \tag{D.14}$$

it is possible to write $Z(\epsilon)$ as a determinant over *spatial* indices of a time ordered product

$$\begin{aligned} Z(\epsilon) &= \det \left[I - \eta T_n \prod_{n=1}^{N_\epsilon} [I - \epsilon h(n\epsilon)] \right]^{-\eta} \\ &\rightarrow \det \left[I - \eta T_\tau \left(\exp \left[-\int_0^\beta d\tau \, h(\tau) \right] \right) \right]^{-\eta} . \end{aligned} \tag{D.15}$$

For time independent Hamiltonians, we can diagonalize $\hat{h}_{ii'}$:

$$h_{ii'} \rightarrow (\epsilon_k - \mu)\delta_{kk'}, \tag{D.16}$$

and recover the free energy of noninteracting bosons or fermions:

$$F = \begin{cases} \beta^{-1} \sum_k \ln \left(1 - e^{-\beta(\epsilon_k - \mu)} \right) & \text{bosons} \\ -\beta^{-1} \sum_k \ln \left(1 + e^{-\beta(\epsilon_k - \mu)} \right) & \text{fermions} \end{cases}, \tag{D.17}$$

which has been derived previously in (A.24).

For finite lattices at finite temperature, $Z[j(\tau)]$ can be computed numerically for arbitrary time dependent source terms. Formulas such as (D.15) are useful, for example, in quantum Monte Carlo simulations of the Hubbard model.[2] The functional $Z[Q(\tau)]$, where $Q(\tau)$ is an auxiliary field that decouples the four fermion interactions, is computed using (D.15). The result yields an effective Boltzmann weight for $Q(\tau)$, which is updated using a Monte Carlo (Metropolis) algorithm. However, one is cautioned that $Z[Q]$ is not positive definite, and its negative signs are highly problematic for these simulations.

---

[2]See Chapter 5 in *The Hubbard Model*, bibliography of Chapter 3.

D.3 Matsubara Representation

For evaluation of dynamic response functions, it is useful to diagonalize $G(\tau, \tau')$ using the Matsubara representation. One defines the Fourier transform into Matsubara representation by the limit of infinite timesteps

$$
\begin{aligned}
X_\omega &= \lim_{\epsilon \to 0} \sum_{\tau=0}^{\beta-\epsilon} e^{-i\omega\tau} X_\tau \epsilon \\
&= \int_0^\beta d\tau \, e^{-i\omega\tau} X(\tau).
\end{aligned}
\tag{D.18}
$$

The inverse transformation is given by the *Matsubara sum*

$$
X(\tau) = \beta^{-1} \sum_{n=-\infty}^{\infty} e^{i\omega_n \tau} X_{\omega_n},
\tag{D.19}
$$

where ω_n are the Bose or Fermi Matsubara frequencies:

$$
\omega_n = \begin{cases} 2n\pi/\beta & \text{Bose} \\ (2n+1)\pi/\beta & \text{Fermi} \end{cases}.
\tag{D.20}
$$

The measure of the path integral can be changed to Matsubara representation

$$
\mathcal{D}^2\mathbf{z} = \prod_{i,\tau} d^2 z_{i,\tau} = \prod_{i,n} \frac{d^2 z_{i,\omega_n}}{\beta},
\tag{D.21}
$$

and the generating functional is given by

$$
Z = \oint \mathcal{D}^2 z \exp\left[-\sum_n \left(-i\beta^{-1}\omega_n \mathbf{z}_\omega^* \mathbf{z}_\omega \right) - \int_0^\beta d\tau H[j] \right].
\tag{D.22}
$$

Using (D.12) and (D.18), the Matsubara representation of the Green's function for a time independent bilinear Hamiltonian is given by

$$
G_0(k, i\omega) = -\frac{\delta_{nn'}\delta_{kk'}}{i\omega - h_k}.
\tag{D.23}
$$

D.4 Matsubara Sums

For complex analytic functions $g(z)$ that are bounded at infinity by

$$
|g(z)| \leq \frac{c}{|z|}, \quad |z| > R_0,
\tag{D.24}
$$

the Matsubara sum for $\tau > 0$ is given by the Cauchy integral formula:

$$
\beta^{-1} \sum_n e^{i\omega_n \tau} g(i\omega_n) = -\eta \sum_\alpha e^{z_\alpha \tau} n(z_\alpha) \, \text{Res}[g(z_\alpha)]
$$

$$
-\frac{\eta}{2\pi i} \sum_\gamma \int_{C_\gamma} dz \; e^{z\tau} n(z) g(z) \,, \quad (D.25)
$$

where $z_\alpha, \text{Res}[g(z_\alpha)]$ and C_γ are the poles, residues, and cuts of $g(z)$ in the complex plane, respectively. $n(z)$ is the Bose or Fermi function

$$
n(z) = \frac{1}{e^{\beta z} - \eta}, \quad (D.26)
$$

which has simple poles at the Matsubara frequencies $z = i\omega_n$. Its residues are

$$
\text{Res}\left[n(i\omega_n)\right] = \frac{\eta}{\beta}. \quad (D.27)
$$

The imaginary time Green's function, for $\tau > 0$, is given by performing the Matsubara sum over (D.23),

$$
G_0(k, \tau) = \beta^{-1} \sum_n e^{i\omega_n \tau} G_0(k, \omega_n)
$$

$$
= \eta e^{h_k \tau} n(h_k) = \eta e^{(\epsilon_k - \mu)\tau} n(\epsilon_k - \mu). \quad (D.28)
$$

Another useful example is the calculation of the dynamical spin susceptibility of a noninteracting system of spin $\frac{1}{2}$ bosons or fermions. The eigenstates are labelled by lattice momentum $\mathbf{k}$ and spin index $s = \pm\frac{1}{2}$. The source terms are written in the Matsubara representation as

$$
\int_0^\beta d\tau \sum_{i,\alpha} j_i^\alpha(\tau) z_{i,s}^*(\tau) \sigma_{ss'}^\alpha z_{i,s'}(\tau)
$$

$$
= -\beta^{-1} \sum_{\mathbf{k}n\mathbf{q}m} j_{\mathbf{q}}^\alpha(\nu_m) \sum_{ss'} z_{\mathbf{k}\omega_n,s}^* \sigma_{ss'}^\alpha z_{\mathbf{k}+\mathbf{q}\omega_n+\nu_m,s'}
$$

$$
\equiv \mathbf{z}^\dagger \hat{j} \mathbf{z}, \quad (D.29)
$$

where

$$
j_{\mathbf{q}}^\alpha(\nu) = \sum_i \int_0^\beta d\tau \, j_i^\alpha(\tau) e^{-i\nu\tau - i\mathbf{q}\cdot\mathbf{x}_i}. \quad (D.30)
$$

Since we choose $j_i^\alpha(\tau)$ to be periodic on the interval $[0, \beta]$, its Matsubara frequencies are of the Bose type:

$$
\nu_m = 2m\pi/\beta. \quad (D.31)
$$

The generating functional is given by

$$
\begin{aligned}
\ln Z[j] &= -\eta \operatorname{Tr} \ln \left[\epsilon (G_0^{-1} - \hat{j}) \right] \\
&= \eta \operatorname{Tr} \ln (\epsilon G_0) + \sum_{n=0}^{\infty} \frac{1}{n} \operatorname{Tr} \left(G_0 \hat{j} \right)^n .
\end{aligned} \tag{D.32}
$$

By (B.19) and (B.24), the susceptibility is given by the second-order term in this expansion. Thus,

$$
\begin{aligned}
\chi^{zz}(\mathbf{q}, i\nu) &= \frac{\partial^2 \ln Z}{\partial j_{\mathbf{q}}(i\nu) \partial j_{-\mathbf{q}}(-i\nu)} \\
&= \frac{\eta}{2\beta} \sum_{\mathbf{k},n} G_0(\mathbf{k}, i\omega_n) G_0(\mathbf{k}+\mathbf{q}, i\omega_n + i\nu) \\
&= -\frac{1}{2} \sum_{\mathbf{k}} \frac{n_{\mathbf{k}} - n_{\mathbf{k}+\mathbf{q}}}{\epsilon_{\mathbf{k}} - \epsilon_{\mathbf{k}+\mathbf{q}} + i\nu},
\end{aligned} \tag{D.33}
$$

where $n_{\mathbf{k}} = n(\epsilon_{\mathbf{k}} - \mu)$. As shown in (B.26) the analytic continuation $i\nu \to \omega + i0^+$ yields the real and imaginary dynamical spin susceptibilities

$$
\begin{aligned}
\operatorname{Re} \chi^{zz}(\mathbf{q}, \omega) &= -\frac{1}{2} \mathcal{P} \sum_{\mathbf{k}} \frac{n_{\mathbf{k}} - n_{\mathbf{k}+\mathbf{q}}}{\epsilon_{\mathbf{k}} - \epsilon_{\mathbf{k}+\mathbf{q}} + \omega}, \\
\operatorname{Im} \chi^{zz}(\mathbf{q}, \omega) &= \frac{\pi}{2} \sum_{\mathbf{k}} (n_{\mathbf{k}} - n_{\mathbf{k}+\mathbf{q}}) \delta(\epsilon_{\mathbf{k}} - \epsilon_{\mathbf{k}+\mathbf{q}} + \omega).
\end{aligned}
$$

$$\tag{D.34}$$

D.5   Exercises

1. The equal time Green's function is defined as the limit

$$
\lim_{\tau \to 0^-} G(kk, \tau) = \langle a_k a_k^\dagger \rangle . \tag{D.35}
$$

Evaluate the Matsubara sum

$$
G_0(k, -\tau) = -\eta \beta^{-1} \sum_n \frac{e^{-i\omega_n \tau}}{i\omega_n - \epsilon_k} \tag{D.36}
$$

by the contour integration method outlined above and verify that

$$
\langle a_k a_k^\dagger - \eta a_k^\dagger a_k \rangle = 1. \tag{D.37}
$$

Hint: The summand in (D.36) is not invariant under $i\omega_n \to -i\omega_n$!

Bibliography

A recommended text for this material is:

- J.W. Negele and H. Orland, *Quantum Many Particle Systems* (Addison-Wesley, 1988).

See also the bibliographies of Chapters 1, 2, and 10.

Appendix E

The Method of Steepest Descents

The method of steepest descents has several names in the literature, such as the *saddle point expansion* and the *stationary phase approximation*. The basic principle is illustrated by the expansion of a contour integral of a single complex variable:

$$I(g) = \int_C dz \, e^{-gf(z)}, \tag{E.1}$$

where $z = x + iy$ runs on the contour C in the complex plane and connects two points where $\mathrm{Re}f \to \infty$.

We assume that $f(z)$ is analytic in the complex plane. The Cauchy–Riemann conditions are

$$\begin{aligned}
\partial_x \mathrm{Re}f &= \partial_y \mathrm{Im}f, \\
\partial_y \mathrm{Re}f &= -\partial_x \mathrm{Im}f,
\end{aligned} \tag{E.2}$$

which also implies that neither the real nor the imaginary part of any analytic function can possess an absolute minimum since

$$\nabla^2 \mathrm{Re}f = \nabla^2 \mathrm{Im}f = 0. \tag{E.3}$$

However, $\mathrm{Re}f(z)$ can acquire a minimum along a contour. Since $e^{-gf(z)}$ is also analytic, we can deform $C \to C'$ in the complex plane and make it pass through the saddle point z_0, which obeys

$$\partial_x \mathrm{Re}f(z_0) = \partial_y \mathrm{Re}f(z_0) = 0. \tag{E.4}$$

C' must obey the following conditions:

1. The endpoints of C' are the same as those of C.

2. $\mathrm{Re}f(z)$ has a *minimum* on the contour at z_0.

3. $\mathrm{Im}f$ is constant along the contour near z_0.

The paths C and C' are illustrated in Fig. E.1.

By (E.2) and (E.4), the third condition is equivalent to the path being parallel to $\nabla \mathrm{Re}f$, i.e., C' is the *path of steepest descents* of $\exp(-g\mathrm{Re}f)$, hence the name of this method.

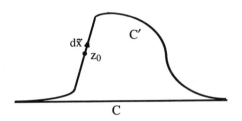

FIGURE E.1. The saddle point and its path of steepest descents.

We determine C' by rewriting our integral as

$$
\begin{aligned}
I(g) &= e^{-gf(z_0)}\, I'(g)\,, \\
I'(g) &= \int_{C'} dz \,\exp\left\{-g[f(z+z_0)-f(z_0)]\right\}\,,
\end{aligned}
\tag{E.5}
$$

$f(z)$ is expanded in a Taylor series about z_0:

$$
f(\tilde{x}+z_0)-f(z_0) = \frac{1}{2}f^{(2)}\tilde{x}^2 + \sum_{n=3}^{\infty}\frac{f^{(n)}}{n!}\tilde{x}^n.
\tag{E.6}
$$

The contour $\tilde{x}\in C'$ is chosen such that

$$
f^{(2)}\tilde{x}^2\Big|_{\tilde{x}\in C'} = |f^{(2)}\tilde{x}^2| \equiv y^2.
\tag{E.7}
$$

C' is (to second order) the path of stationary phase,

$$
\mathrm{Im}f(\tilde{x}+z_0) = \mathrm{Im}f(z_0) + \mathcal{O}(\tilde{x}^3).
\tag{E.8}
$$

I' is expanded in powers of g^{-1} as follows:

$$
\begin{aligned}
I'(g) = {}& \sqrt{\frac{2\pi}{gf^{(2)}}} \int_{-\infty}^{\infty}\frac{dy}{\sqrt{2\pi}}\, e^{-\frac{1}{2}y^2} \\
&\times \sum_{m=0}^{\infty}\frac{1}{m!}\left(-\sum_{n=3}^{\infty}\frac{f^{(n)}}{n!g^{\frac{n}{2}-1}(f^{(2)})^{\frac{n}{2}}}y^n\right)^{m}.
\end{aligned}
\tag{E.9}
$$

All odd powers of y vanish by symmetry. The integrand is rewritten as a power series in y^{2n} with coefficients a_{2n}. The Gaussian integrals are given by

$$\int_{-\infty}^{\infty} \frac{dy}{\sqrt{2\pi}} y^{2n} e^{-\frac{1}{2}y^2} = 1 \cdot 3 \cdot 5 \ldots (2n-1) = (2n-1)!!, \qquad (E.10)$$

which yields

$$I'(g) = \sqrt{\frac{2\pi}{gf^{(2)}}} \left[1 + \sum_{n=1}^{\infty} g^{-n} a_{2n} (2n-1)!! \right]. \qquad (E.11)$$

There are cases where multiple saddle points exist,

$$\partial_x \mathrm{Re} f \bigg|_{z_0^\alpha} = 0, \quad \alpha = 1, 2, \ldots \quad . \qquad (E.12)$$

If the distance between saddle points is much larger than the typical fluctuations in I',

$$|z_0^\alpha - z_0^{\alpha'}| >> \frac{1}{\sqrt{gf^{(2)}}}, \qquad (E.13)$$

one can sum over saddle points as independent integrals. This yields

$$I(g) = \sum_\alpha e^{-gf(z_\alpha)} I'_\alpha(g), \qquad (E.14)$$

where I'_α is the fluctuation integral about z_0^α.

The generalization of the steepest descents expansion to a multidimensional integral is straightforward. Given an integral of the form

$$I(g) = \int \mathcal{D}\mathbf{z} \, e^{-gf(\mathbf{z})}, \qquad (E.15)$$

where $\mathbf{z} = (z_1, \ldots, z_N)$, we search for the saddle point

$$\partial_{x_i} \mathrm{Re} f \bigg|_{\mathbf{z}_0} = 0, \quad i = 1, \ldots, N. \qquad (E.16)$$

The saddle point $\mathbf{z}_0$ is, in general, a complex vector, which minimizes $\mathrm{Re} f(\mathbf{z})$ along the N contours parametrized by x_i. The contours are chosen such that $\mathrm{Im} f$ is constant near the saddle point. Equations (E.16) are N coupled nonlinear equations. In the literature they have many names: "saddle point equations," "classical equations of motion," "mean field equations," etc., depending on the integral in question and the large parameter that controls its expansion.

The quadratic form of f about a saddle point $\mathbf{z}_0$ defines the inverse "propagator" D:

$$D_{i,j} = \frac{1}{2} \left(\partial_{\tilde{x}_i} \partial_{\tilde{x}_j} f \right)^{-1}_{i,j}. \tag{E.17}$$

D generalizes the inverse of the Gaussian coefficient $f^{(2)}$. It describes effective interactions between the fluctuations in the variables $\vec{\tilde{x}}$. It is often called in the literature the *"RPA propagator."*[1]

Bibliography

A short and clear reference for the method of steepest descents can be found in

- G. Arfken, *Mathematical Methods For Physicists* (Academic Press, 1970), Section 7.4.

---

[1] which stands for "random phase approximation." See, e.g., Section 17.2.

Index